ICE

Ice

Fred Hoyle

Hutchinson

London Melbourne Sydney Auckland Johannesburg

Hutchinson & Co. (Publishers) Ltd
An imprint of the Hutchinson Publishing Group
3 Fitzroy Square, London W1P 6JD

Hutchinson Group (Australia) Pty Ltd
30–32 Cremorne Street, Richmond South, Victoria 3121
PO Box 151, Broadway, New South Wales 2007

Hutchinson Group (NZ) Ltd
32–34 View Road, PO Box 40–086, Glenfield, Auckland 10

Hutchinson Group (SA) (Pty) Ltd
PO Box 337, Bergvlei 2012, South Africa

First published 1981
© Fred Hoyle 1981

Set in Times
by D. P. Media Limited, Hitchin, Hertfordshire

Printed in Great Britain by
The Anchor Press Ltd and bound by
Wm Brendon & Son Ltd, both of
Tiptree, Essex

British Library Cataloguing in Publication Data

Hoyle, Fred
  Ice.
  1. Climatic changes
  I. Title
  551.6'4'23   QC981.8.C5

ISBN 0 09 145320 8

For Elizabeth

Contents

Illustrations

Tables

Acknowledgements

It is a pleasure to thank Dr Clark Friend for the care with which he has been over the book, for removing errors, and for giving me fair warning when I am on dangerous ground, as in the topic of wind-blown debris.

Introduction

Forty-five years, almost two-thirds of a lifetime ago, I was a research student in the University of Cambridge. There were not too many of us 'doing' research in those days, and it was one of the virtues of small numbers that we were on arguing terms with each other, not just in our own expertise as it happened to be, but over the sciences at large, and even in such apparently distant matters as history and the classics. It was in those days that my interest in ice ages was born, an interest that has never left me in subsequent years of wandering the glaciated hills of the British Isles.

The ice ages of the most recent geological era, the Cenozoic, were thought to have been confined to just four episodes, well coordinated in the northern hemisphere; but the situation in the southern hemisphere was still uncertain — nobody yet knew that ice ages occur contemporaneously in both geographical hemispheres. It was a curiosity that, in the far more distant past, about 250 million years ago, there had been glaciers in the subcontinent of India. Nothing much could be made of this fact, nor of the coal measures found as far north as Spitzbergen, or of the evidence that a temperate flora had once existed in Antarctica. Although these findings gave support to Alfred Wegener's theory of drifting continents, this theory was not considered respectable in Cambridge. I was lucky, however, in having a particular friend, Joe Jennings, who always spoke up for Wegener.

Three theories of the ice ages themselves were widely discussed. The astronomical theory of Croll and Milankovich, discussed later in Chapter 4, was one of them. The total amount of sunlight considered to reach the Earth's surface does not change in their theory. What does change, but only by a few per cent, is the distribution of sunlight between the northern and southern hemispheres and between the equatorial and polar regions. Since meteorological processes in the atmosphere, by transferring energy from hotter to cooler regions, smooth inequalities of the amounts of sunlight received in the different geographic zones, changing the inequalities by a mere per cent or two could scarcely be of consequence. The Croll–Milankovitch theory was therefore largely discounted in my student days. Although the theory has been revived strongly in recent years, my own opinion has not shifted. For me, the theory lacks sensible physical content, as I shall explain more fully in Chapter 4.

A certain class of theories argues that significant effects can arise from causes too weak to be specified, and in some cases this is true. Let me give an example. If water is boiled in an open pan, the boiling will be observed to start with the rising of streams of bubbles from the bottom of the pan. Suppose at a particular moment one were to take a snapshot of the bubbles. On examining the resulting picture one might ask why there are bubbles in some places but not in others. No answer could be given, because the causes are too small for them to be identified. A theory of this type for the ice ages was widely discussed in the 1930s. It went as follows.

Every ten years or so we experience a severe winter followed by a cool summer. If this pattern were to be repeated year after year for a century or so, winter snow would accumulate in some places that are at present without glaciers, particularly on higher ground. The accumulation of snow might then modify the climate so as to accentuate the pattern, causing the severe winters and cool summers to persist for a further span of years. This in turn could accentuate the situation still more. So, step by step, coming from an original more or less chance clustering of a number of abnormal years, we would be led into an accelerating ice age. If the 'no-cause' theory of the ice ages were correct, the Earth (once in an ice age) should remain ice-bound indefinitely. But the Earth did not remain in the ice ages of the Cenozoic. Indeed the glaciers of the most recent ice age melted suddenly and decisively away in not more than a few centuries.

The no-cause theory is also disproved by a direct experiment, for in medieval times there was indeed a run of abnormal winters and summers lasting for about a hundred years that produced an expansion of northern-hemisphere glaciers, in what is known as the Little Ice Age. Although apparently on course towards a major new ice age, the Earth returned to warmer conditions – the threat did not materialize.

If the Sun becomes weaker in its emission of light and heat the Earth will cool. The third theory of my student days was that the Sun had been weaker during the ice ages. But was this presumption really plausible? With the development of the study of stars by Eddington in the 1920s, and the development of nuclear physics in the 1930s, it could be seen that the idea was quite *im*plausible. Although in the last few years astrophysicists have had worries about the Sun, to a point where all earlier investigations have been carefully re-examined, the situation in this respect remains where it was. While the Sun changes its emission slowly over many hundreds of millions of years, it cannot have varied repeatedly up and down over the shorter timescale covered by the ice ages of the Cenozoic. Thus the theories as they presented themselves forty years ago would seem all three to be wrong.

My first attempt at the ice-age problem, many years ago, came out of the third theory. True, the Sun alone could not vary up and down by significant amounts in only a million years or so, but in the real world the Sun is not alone. It is one star among 100,000 million others in our galaxy. Between the stars there are a number of irregular gas clouds – about 4000 of them, according to recent estimates. From time to time the Sun, together with its retinue of planets, must enter such a cloud. In 1939, R. A. Lyttleton and I calculated that such an encounter would occur about every 250 million years and would last when it occurred for several million years. Could the ice ages of the Cenozoic be connected with irregularities in a recent cloud encounter, and could the glaciation we knew of in India be connected with a previous encounter? But, as observational evidence accumulated against there being a cloud of gas close enough to the solar system for a recent encounter, our idea had also to be abandoned.

After this early setback, it was many years before I returned seriously to the ice-age problem. By then the situation had become ominous. So long as one thought the number of ice ages to have occurred throughout the Earth's history to be no more than about half a dozen, as we used to think in the 1930s, it seemed possible that the recent ice age which terminated about 10,000 years ago might prove to be the end of the sequence of Cenozoic glaciations. It was again my friend Joe Jennings who drew my attention to the error of this belief. My respect for Joe's opinion was such that I began combing over the further evidence now available from the 1960s and 1970s. I found things just as Joe had said. The next ice age is not so much a possibility as a certainty.

With this realization, it was natural to wonder if any planned human action could prevent such a calamity. Since this question depended for its answer on an understanding of the cause of ice ages I returned again to the problem. The situation remained obscure to me until the day I chanced on a description of a quite simple experiment. If air that has not been thoroughly dried, that contains a number of very small water drops, is cooled progressively in a chamber, the droplets do not solidify into ice crystals as their temperature falls below the normal freezing point of water at 0°C. The droplets remain as a supercooled liquid down to a remarkably low temperature, close to −40°C, when at last the liquid water goes into ice. If a beam of light passes through the chamber, and if one looks into it from a direction at right angles to the beam, the chamber appears dark so long as the droplets stay liquid. Their transition to ice is signalled by a sudden radiance from the interior of the chamber, which means that, whereas liquid droplets transmit the light beam, the ice crystals scatter it. The relevance of this observation to the ice-age problem comes from the possibility of ice crystals forming in large numbers from water droplets high in the Earth's atmosphere. An increased fraction of sunlight would then be reflected back into space, with an effect similar to a lowering of the intrinsic emission of the Sun, and so producing an outcome like that of the third theory discussed above. The details will emerge as we proceed through the book.

Besides the main issue of the cause of ice ages there are a number of intriguing ancillary problems. Geologists may be unsympathetic to my arguments concerning the puzzle of the transport of erratic boulders. Even so, I stick to my heterodox view that glaciers and ice sheets cannot spread enormous horizontal distances as they are supposed to have done. Ice accumulated in the English Lake District could not have flowed into central England. Nor could ice which accumulated at the Norwegian coast have flowed to Durham and Yorkshire or to the north German plain or into western Russia.

There is another way that an ice sheet can spread over a large horizontal distance; a continuing deposition of new ice at the boundary of the sheet would spread the sheet without any individual piece of ice moving distances of more than a few kilometres. It is not hard to imagine a sweeping process analogous to the action of a domestic broom whereby the ice at the sheet boundary could push a mass of unsorted rock debris ahead of it. In such a process, the debris could also travel much greater distances than any individual piece of ice.

Erratic material could also be moved through large horizontal distances by ice-rafting. Icebergs containing rock debris float in some cases by thousands of kilometres from their present-day sources in Greenland and Antarctica. When the icebergs eventually melt, the debris simply falls to the ocean bed, which today happens to be mostly in deep water, but which in the ice ages could frequently have been in temporarily flooded areas of shallow water adjacent to the British Isles and Scandinavia.

Yet neither ice-rafting nor broom-sweeping could transport rock debris through considerable heights, and some erratic boulders have been found many hundreds of metres above their sources. These cases require a quite different method of transport to have been operative. If my attempt to find such a different process does not commend itself, I am happy to leave it to the reader to find a better one!

1 The Ice-Age World

The word *Eiszeit* or 'ice age' is said to have been used first by Johann Wolfgang von Goethe, the great German writer. It is a word well suited to describe local climate, for there is a sharp contrast between a particular region being available for the pastoral and arable pursuits of man and the same region being buried under 800 metres of ice. Yet on a world-wide scale the word *Eiszeit* is perhaps not so useful; as long as there are ice sheets anywhere in the world, the Earth may be said to be in an ice epoch. We should rather think in terms of 'more ice' and 'less ice'.

There is still a lot of it today, in a great sheet covering most of the Antarctic. It has been estimated that the Antarctic sheet consists of between 26 and 30 million cubic kilometres of ice. There are perhaps a further 5 million cubic kilometres of ice in Greenland, bringing the current world total to between 30 and 35 million cubic kilometres. If all that ice were melted, the resulting water would raise sea level by perhaps 80 metres, flooding most of the world's major cities.

At the height of the last ice age, 20,000 years ago, there was obviously more ice piled on the land than there is today. As we know that sea level then was 100 metres lower than it is now, it is easy to calculate just how much more there was. The total area of the world's ocean is about 360 million square kilometres, which means that 36 million cubic kilometres of water must have evaporated to provide the extra ice. Since ice has about nine-tenths of the density of water, such a volume of water would have made about 40 million cubic kilometres of ice – that is, about one and a quarter times the amount there is today.

A good deal of the extra ice was added to the Antarctic plateau, but the rest was deposited in the northern hemisphere, wiping out of human use much of the area now occupied by

America, Europe and the USSR. The overall distribution of the ice sheet is shown in Figure 1. It was about 800 metres thick on average and covered such cities as Chicago, Boston, Glasgow, Stockholm, Leningrad and Moscow. Over northern Canada and the mountains of Scandinavia it was thicker still, with a depth of about 2500 metres, comparable to the ice sheets in Greenland and the Antarctic today.

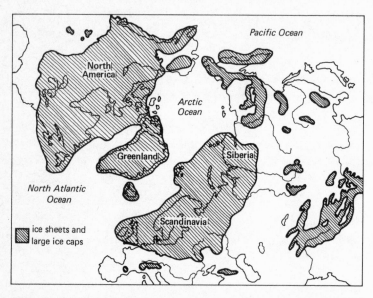

Figure 1: *Ice sheets in the northern hemisphere at 20,000 BP, typical of the distribution of the northern ice during an ice age. There was also sea ice in the Arctic Ocean, much more ice in Antarctica, and even ice on tropical mountains which are ice-free now.*

Recent work on cores of oceanic sediments (see Technical Note 3) has proved that the southern hemisphere did not escape from the severe cold that paralysed the north 20,000 years ago. Evidently the chill of the last ice age affected the whole world. Even the tropics were involved. There is evidence of glaciers on mountains that are now completely free of ice – mountains such as Mauna Kea and Mauna Loa in Hawaii and Mount Elgon in

Uganda. Twenty thousand years ago (or 20,000 years BP, 'BP' standing for before present) tropical mountain glaciers descended to about 10,000 feet (about 3000 metres) above sea level, although nowadays such glaciers only appear higher than 16,000 feet (about 5000 metres) above sea level.

In Great Britain, Scotland, Northern Ireland and the north of England were covered by several hundred metres of ice. Conditions in the ice-free regions shown in Figure 2 were inhospitable – similar to the climate of central Alaska today. The Finnish composer Jean Sibelius wrote a tone poem, *Tapiola*, which describes in sound the dark northern forests of Scandinavia. The music suggests that they are places almost outside human experience: hostile and implacable in winter, breathtakingly beautiful in summer. Yet the forests end at northern limits in Canada as well as Scandinavia and Siberia, giving place to arctic tundra – open land that is even more hostile, implacable and colder than the forests themselves. That is just what the ice-free parts of England were like 20,000 BP. The British climate was too cold for any trees to grow at all. The few plant species which survived even in the less harsh regions of the south were predominantly those that are now found in salty soils. This suggests that the unglaciated parts of Britain were not only very cold, but also very dry, probably with an annual precipitation of less than 10 centimetres water equivalent. (If precipitation occurs as snow rather than rain, the water equivalent is the amount of water that would be yielded if the snow were melted.)

A phenomenon known as ice-wedging, which occurred in East Anglia and central England, also indicates a low, desert-like precipitation rate. Ice-wedging arises in the following way. Regions exposed to an annual average temperature of 1 or 2 degrees below freezing develop a condition known as permafrost, when water in the ground becomes frozen down to a considerable depth below the surface of the land. At a still lower temperature, the ice below the surface sometimes cracks open, forming a narrow rift. In summer the rift fills with water, which then freezes next winter and cracks again at the original line of weakness, causing the initially narrow rift to widen progressively until a considerable wedge has been opened up. As well as severe cold, low winter snowfall is an essential condition in the growth of

19

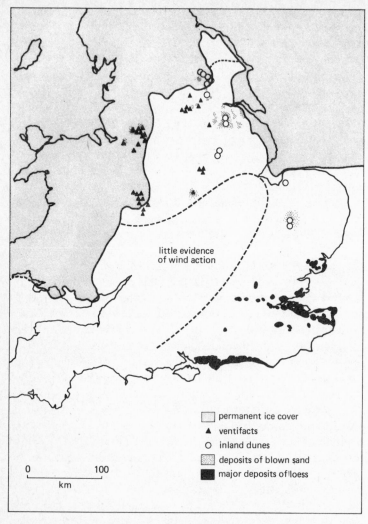

Figure 2: *Glaciated Britain at 20,000 BP. Wind-blown deposits, ventifacts and loess in periglacial regions indicate a dry climate, like that in a cold desert. (After R. B. G. Williams in A. E. Wright and F. Moseley (eds.),* Ice Ages: Ancient and Modern, *Seel House Press, 1975).*

ice-wedges. Wedges will not crack if there is a thick snow cover because snow is an excellent heat insulator. In an example quoted in the *Encyclopaedia Britannica*, the temperature under 30 centimetres of snow was only $-1°C$, although the surrounding air temperature was $-18°C$. This explains why the mountaineers caught in severe winter storms who survive are those who dig themselves into snow holes.

Studies have shown that ice-wedging does not happen unless the average annual temperature is less than $-7°C$, as for example, in parts of Alaska and Siberia today. By combining this information with the knowledge that there were no trees in southern England 20,000 years ago, it is possible to estimate the annual temperature. Figure 3 shows an estimated annual temperature curve (curve B) for ice-age Britain contrasted with the present monthly average temperature of the English Midlands (curve A). The summer difference between curve A and curve B is much less extreme than the winter difference; in the terminology of the old Fahrenheit temperature scale, winters, according to curve B, had an average temperature of '45° of frost'.

Winds seem to have been rather light. While wind-blown ice-age material has been found in southern Britain, it is thin and patchy, unlike the deposits that have been found in eastern Europe, North America and China. The only substantial accumulation of wind-blown material in Britain is to be found in the Breckland, a region about 20 miles (30 kilometres) wide which straddles the Cambridgeshire–Norfolk border.

Although we cannot be certain how men close to the great ice sheets of the northern hemisphere lived, it is possible to make a fair guess. We know how the Eskimos of Alaska lived before the coming of the white man, and there cannot be much scope for variation in the conditions of survival between recent Alaska and the Europe of 20,000 BP.

Survival must have depended on a highly technical society. There would be spears; harpoons for those living near rivers and the sea; sledges, bows and arrows and bone implements in general. There would be skins for clothing – perhaps two sets, the inner set with its fur inwards and the outer with its fur outwards. Although ice-age man's technology was apparently much simpler than ours, it was just as crucial to his way of life. Ice-age man was already a technocrat, easily able when in a later age circumstances changed to smelt steel and to design modern electronic equipment.

The concept of Eskimos living in shelters built from snow, the so-called 'igloos', is exaggerated. Eskimos used whatever shelter they could find, and this must surely have been true for the

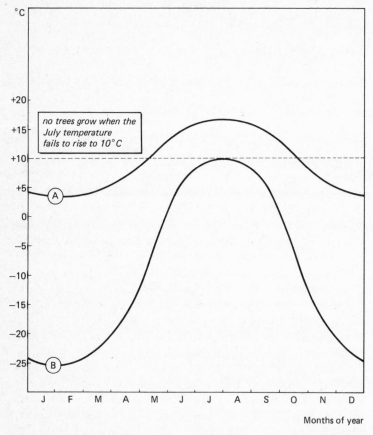

Figure 3: *Temperature curves for central England: curve A for the present day; curve B for estimated ice-age temperatures.*

22

people of 20,000 BP. In Europe, they lived in suitable caves. Doubtless, tents made from skins were also employed.

Communities were probably based on kinship – 'kinship' having a much broader meaning that blood relationship. On marriage, a man acquired his wife's relations as kin, even though there was no blood relationship. Your kinsfolk were the people on whom you might count for a favour. Perhaps even more important, your kinsfolk were the people on whom you might depend to do you no harm.

Hunting parties would be made up as the occasion warranted. For small game, the men would hunt either alone or in twos and threes; but, for the pursuit of large animals, the whole of a community would band together.

While two or more communities might join in a cooperative venture, there were probably no set rules by which they did so. From time to time, a man of outstanding personality and ability would emerge, and several communities might elect to follow his leadership for a while. But there would be no nations in the modern sense, no large clans dedicated to the defence of particular territories. Those of us in modern times who flinch under the lash of bureaucratic rules and who find customs and passports irksome, have ice-age ancestors among our forbears. Another thing too, our ice-age ancestors were surely full of jokes and laughter. This is so with the Eskimo, and it is so with modern parties of mountaineers. Under really hard conditions there is no room for the moody and morose.

The presence of ice sheets in the Antarctic and Greenland today shows that we are still in an ice epoch, even though we are enjoying a comparatively warm 'interglacial' period. Figures 4 and 5 show that, over the past 2300 million years, there have been several ice epochs. Even if the current one, marked 'Cenozoic' in Figure 4, is not destined to last as long as the 100 million years of the great Permo-Carboniferous ice epoch, it is nonsense to suggest that the present one has ended and that there will never be any more ice on the Earth than there is now. On the contrary there is every reason to assume that our own Cenozoic sequence of ice ages has barely begun. It is clear from Figure 6 that there have been many oscillations of climate in north-

23

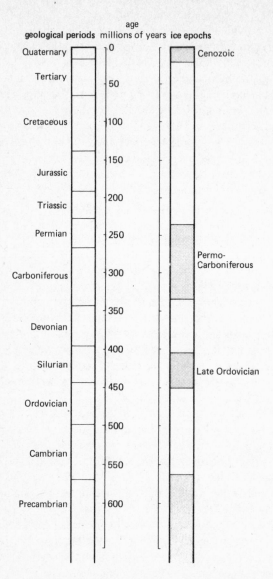

Figure 4: *Geological periods* (left) *and durations of ice ages* (right).

western Europe, ranging from the upper, warm level of the present day to the lower, exceedingly harsh, level of the ice ages. Figure 6 also shows that the variations of climate have been growing more extreme and rapid in their oscillations. Considering the curve of Figure 6, I would say that there is no chance of avoiding another ice age, unless we take deliberate action to prevent it.

Recent work has shown that the fluctuations between warm and cold conditions have occurred with astonishing swiftness. Figure 7 is a temperature chart for the periglacial regions of Great Britain from 50,000 BP to 17,000 BP. The chart was calculated from a study of British beetles carried out by G. R.

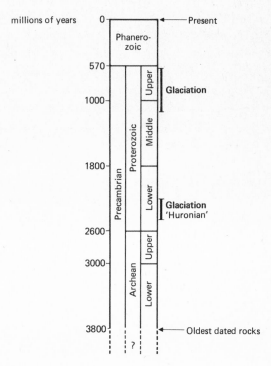

Figure 5: *Details of the Precambrian period.*

25

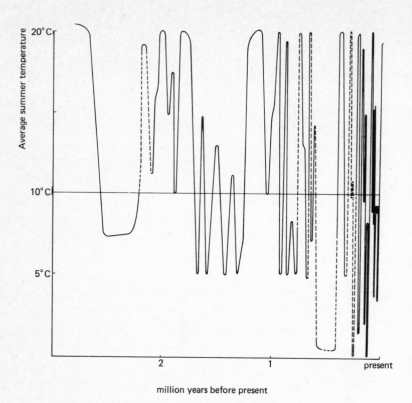

Figure 6: *Recent ice ages. An ice age occurs when the average summer temperature falls below 10°C. (Adapted from W. H. Zagwijn in A. E. Wright and F. Moseley (eds.),* Ice Ages: Ancient and Modern, *Seel House Press 1975).*

Coope and his collaborators.[1] The skeleton of a beetle is made from hardened chitin (the material of crab shells), which is very tough and survives almost perfectly when buried in organic material and general detritus. In some cases, even the wings, antennae and limbs of the beetles have survived almost undamaged. This means that it is possible to identify precisely which beetles

[1] *Ice-Ages: Ancient and Modern*, A. E. Wright and F. Moseley (eds.), Seel House Press, Liverpool, 1975.

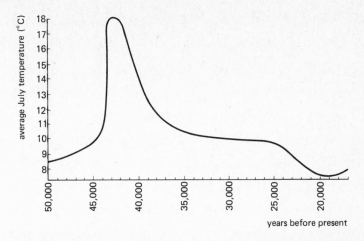

Figure 7: *Average summer temperatures in southern England from 50,000 to 17,000 BP. (After G. R. Coope in A. E. Wright and F. Moseley (eds.),* Ice Ages: Ancient and Modern, *Seel House Press, 1975)*

are present in the deposits of various ages. Since the climatic range during which each species of beetle flourished is known, the combination of species found in the fossil record provides a rather precise definition of the climate – especially the summer temperature – that existed when they were deposited. The radiocarbon technique described in Technical Note 1 gives the age of the deposit; using this, it is possible to construct an accurate temperature chart of the past for the British Isles.

The significance of Figures 6, 7 and 8 is dramatic. Figure 7 shows that, about 43,000 years ago, there was a short-lived interlude of 2000 or 3000 years when summer temperatures were as high as they are now. But around 40,000 BP the climate deteriorated rapidly, with the average July temperature falling close to 10°C, as shown by curve B of Figure 3. During the coldest time of the last ice age – about 20,000 BP – the July temperature fell below 8°C.

The sudden warming of some 43,000 years ago was remarkable for the rapidity of its occurrence which was even swifter than the subsequent reversion to glacial conditions. A similar sudden

27

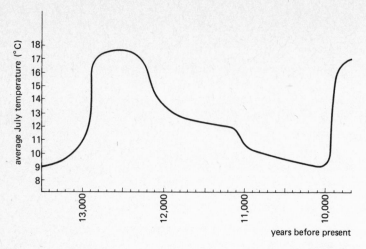

Figure 8: *Average summer temperatures in southern England from 13,500 to 9700 BP. (After G. R. Coope in A. E. Wright and F. Moseley (eds.),* Ice Ages: Ancient and Modern, *Seel House Press, 1975.)*

pulse of warmer climate occurred more recently – about 12,500 years ago – shown in Figure 8. The emergence from cold to warm conditions took only a few decades, while the return from warm to cold took a century or two at most. In a recent article, G. Woillard remarked on an even more rapid catastrophic cooling that took place some 115,000 years ago in north-western Europe in less than two decades.[2]

Clearly, when the next ice age comes, it will come quickly. And, when it comes, it will plunge us back into the conditions survived by our ancestors – conditions that will be disastrous for our present-day civilization.

[2] G. Woillard, *Nature*, vol. 281 (1979), p. 558.

2 Ice Age Makes Man

It is ironic that it was the enormous changes in world ecology caused by earlier ice ages that gave man the chance to become the dominant animal on Earth.

Evidence derived from the buried pollen of plants which lived in past ages (see Technical Note 2) suggests that the earth has been slowly cooling over a timescale of many millions of years. As a consequence of this progressive cooling, which is shown schematically in Figure 9, the first small glaciers were formed in Antarctica perhaps as long ago as 40 million BP. They expanded gradually until, about 20 million BP, a permanent ice sheet

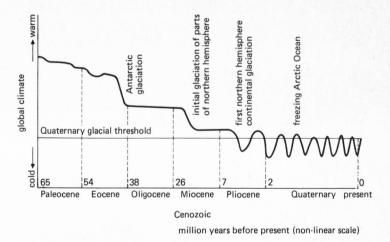

Figure 9: *Global climate over the last 65 million years. The Earth has cooled gradually. An ice-age 'threshold' set in about 3 million years ago. (After J. Andrews in B. S. John (ed.),* Winters of the World, *David and Charles, 1979.)*

29

covered the whole Antarctic continent. About 10 million years later, glaciers appeared on the high mountains of Alaska, and about 3 million years ago, ice sheets developed on lower ground in high northerly latitudes.

It was during the later stages of this cooling of the Earth that our ape-like ancestors first emerged from the forests to try survival on the neighbouring grasslands – a venture reminiscent of the still greater experiment some 340 million years ago when a remote amphibian ancestor of ours climbed out of the sea to adopt a way of life on the land. Animals rarely make a drastic change in their way of life unless forced to do so by circumstances such as a reduction in food supplies. The cooling climate probably caused the forest zone to shrink towards the equator and, as the forest contracted, the competition for survival among its inhabitants would have intensified. For some time our ancestors must, I think, have spent much of their time foraging for food on the ground, rather as the modern gorilla does today. As the area of the forest floor became smaller, the fight for its somewhat scanty pickings would have become more and more difficult, until eventually our ancestors gave up the battle and burst out into the open grasslands.

Our ape-men forefathers had no obvious natural weapons in the struggle for survival in the open. They had no significant armour for attack or defence, no great defensive bulk like the elephant, nor tough skin like the rhinoceros. They had neither the powerful teeth nor the strong claws of the big cats. They could not compare with the bear, whose strength, speed and manipulative claws provided an impressive 'small-fire' weaponry. They could not even defend themselves by running swiftly like the horses, zebras or small animals. If the ape-men had attempted to compete on those terms in the open, they would have been doomed to failure and extinction. But they were endowed with enormous concealed advantages of a kind not possessed by any of their competitors.

In the search for the pickings of the forest, the ape-men had developed efficient stereoscopic vision and a sense of colour that the animals of the grasslands did not possess. The ability to see clearly at close range permitted the ape-men to study practical problems in a way that lay far beyond the reach of the original inhabitants of the grasslands. (No descendant of theirs would ever be able to read a book, conduct a laboratory experiment or attend surgically to a wound.) Good long-distance sight was quite another matter. Lack of long-distance vision had not been a problem for forest-dwelling apes and monkeys because the higher the viewpoint, the greater the range of sight – so all they had had to do was climb a tree. Out in the open, however, this simple solution was not available. Climbing a hill would have helped – our modern instinct to climb to the top of eminences may well be a distant memory of earlier necessity – but in many places the ground was flat. The ape-men adopted the only possible solution. They reared up as high as possible on their hind limbs and began to walk upright.

This vital change of posture entailed considerable disadvantages. It was extremely unstable and it meant that the already slow ape-men became slower still; indeed, on rough, uneven ground they would have been reduced to a pitiful crawl. However, they persevered and their bone structure gradually became adapted to the new, unstable position that earned them the name *Homo erectus*, upright man.

The upright men's good colour vision was useful for the gathering of berries and fungi, but those did not provide sufficient food for survival. More food of higher quality was necessary if *Homo erectus* was not to die out. The most obvious source of nourishment, the grass, was useless because his stomach was not adapted to digest it. Meat was the obvious answer, but to obtain it he had to compete directly with powerful animals of prey that had evolved both armoury and great speed over many millions of years. He had just one physical advantage – his hands. Hands which had evolved in the forest to grip the boughs of trees could now grip sticks. A modern man equipped with a heavy blackthorn stick could probably get the better of a dog, although probably not of a wild cat, whose greater speed would allow it to dodge and attack successfully from behind. The solution would be for a party of men to form themselves into a group that could ward off attacks from all directions. The new upright men had therefore to cooperate with each other to survive, and they had to exercise just the kind of physical discipline that is used by modern armies in the field.

31

How effective could such an apparently primitive deployment of forces have been? I suspect that a group of vigorous men armed with heavy sticks would be more than a match for an equal number of hyenas, and it is now known that a pack of hyenas can compete effectively with a pride of lions. Besides which, it is, I think, true that no modern wild animal will risk an encounter in the open with just such a group of men. There seems to be an instinctive feeling in other animals that a stick is an exceedingly dangerous weapon. We ourselves have attached a strange mystic significance to a symbolic stick, the sceptre, and to the power that is supposed to go with its possession. The implication is that, armed with sticks, *Homo erectus* was able to compete.

All this sounds rather straightforward, but it actually conceals a quite subtle mathematical problem. Cooperation is vital, but the larger the number of individuals in a group, the more food the group needs to sustain itself. The group would gain ground if the extra food obtained through the addition of a new individual exceeded the amount of food required by that individual. But, once the individuals' requirements exceeded their contributions, the group would lose ground with each new individual added.

Under primitive conditions, small groups tend to gain from the addition of new individuals and to build up to a balancing number at which food cost equals food gain. If the species is to survive, the optimum hunting group must be able to compete successfully with groups of other species, or extinction is inevitable.

The tactics used by the early humanoids must at first have been strange to the animals of the grasslands and, because of the novelty, the newcomers would have been successful enough at first to build optimum groups of a fair size, perhaps as large as the hunting packs of modern hyenas. There is no standstill, however, in the interplay between prey and predator. Prey is always finding new ways to frustrate its predator, who must then develop more effective hunting tactics in order to survive. There were two ways in which humanoids could improve their tactics: by more efficient cooperation and by inventing better weapons.

Improved cooperation must have required the passing of accurate information from one individual to another with fewer misunderstandings in moments of stress. These two requirements are different. Information can be passed in moments of leisure by signs and facial gestures, but there would be no time for hand signs in the tense moments of a hunt. Vocal signals would have had to be used and, for accuracy, precisely controlled repetitive sounds would have been essential. I believe that language had its origin in violent action rather than in leisurely communication, and I think one could detail many aspects of language that support this belief. The effect of a compulsive stutterer on a roomful of people is remarkable in its emotional intensity – quite unlike the calm way in which we look on a limp from a broken leg. Information from a stutterer would scarcely be of much use if one were seeking to beat a pride of lions away from its kill. Of course, someone with a broken leg would be equally useless, but an injured man would not be one of a hunting party.

The need for better weapons would have become clear to the upright men as they saw their prey escaping from them. Defensive animals do not run enormous distances from their predators but only the minimum distance that experience has shown to be necessary – if the wildebeest were to run 50 miles from a lioness, it would lose a great deal of energy and would probably have moved out of its favourite feeding grounds. Because the humanoids moved so slowly, the minimum distance necessary for their prey to run from them was very small. Swift animals could stop a few steps out of range of the sticks, tantalizing the hungry men. In time, the men learnt to pick up stones and throw them.

We do not think of stones as effective weapons nowadays because our arms are not properly developed. But the power with which major baseball players in the United States can throw a ball around the infield suggests just how formidable a weapon a stone must have been.[1] At first, the weight of the stone – and the strength of its impact – would have seemed all-important, but there must have been occasions when an animal managed to escape in spite of being hit. Gradually, the humanoids must have learnt that such failures were less likely to occur if the stone had sharp edges. They would have started by searching for natural stones of the most effective shape until some experimental

---

[1] I am indebted to Dr Clark Friend for pointing out that Eskimo children in Greenland throw stones at anything which moves, and that, like baseball players, they become experts at throwing.

genius among them had the idea of deliberately chipping a stone to produce the required edge. The stones could then be lighter and thrown from further away. Chipped pieces of stone could even be attached to sticks to form spears which could be thrown further still.

Increasing the range of weapons never ceased to be important, because the defensive animals learnt to increase the safe distance from the hunters. Eventually, that distance became even greater than the range of a lightweight stone or spear. Although the invention of the bow and arrow continued to extend the hunters' range of attack, effective hunting came in the end to depend overwhelmingly on guile, and in this our ancestors were able to deploy their intellectual advantages over the other inhabitants of the grasslands.

All the grassland predators could use concealment as a tactic, and they learnt to divide themselves into more than one party to baffle their prey. What *Homo erectus* and his descendants could do in addition to this was to prepare ambushes and dig traps to snare even very large and powerful animals such as the elephant, tiger, bear and woolly mammoth.

After causing the reduction of the forests, the developing global ice ages did not play any very great role in the evolutionary development described to this point. But the increasing cold had important effects in other ways. There were woodland areas dotted over the grasslands, and occasionally lightning started a fire in one of these areas. It is easy to imagine a group of fifty to a hundred men, women and children in a bitter winter long ago rushing to seek the warmth of an apparently miraculous fire. A genius in such a group, probably about a million years ago, not only perceived the enormous advantages of an artificially induced fire, but must have set about the task of discovering how to light it. The individual or individuals who first managed to produce fire from tinder made an invention of the highest order.[2]

Over millions of years, the unending struggle for survival, for food and warmth, the struggle to solve one practical problem after another, forced the human brain to develop in size and

[2] For possible methods see *A History of Technology*, vol. 1, Oxford University Press, 1958.

complexity. The cranial capacity eventually grew to about 1600 cubic centimetres from about half that volume.

Homo erectus branched into species and subspecies along an evolutionary tree of the usual complex form, and this led to competition between different branches of that tree. The pressures were unremitting, especially on those who sought to maintain themselves in the periglacial lands of the northern hemisphere. Outstanding among the subspecies was *Homo neanderthalis*, who took up residence in Europe at about 125,000 BP. The name of the subspecies was taken from the gorge-like Neander Valley, about 10 kilometres east of Düsseldorf in Germany, where the first skeleton was found in 1856. Two more skeletons were found thirty years later in the Cave of Spy in Belgium, and from a close examination of all these remains scientists have established that Neanderthal man was like, but not identical to, modern man.

The early reconstructions, however, contained errors which led to the belief that Neanderthal man was a stooping, shambling creature, and modern popular journalism (the tide of which one can do nothing to halt) has broadcast the quite false idea that Neanderthal man is to be equated with primitive man. There is no evidence that Neanderthal man was in any way inferior to ourselves and, given that he survived in Europe throughout the long harsh period from about 125,000 BP to about 35,000 BP, he must have been both extremely hardy and highly intelligent. Ironically, Neanderthal man was destroyed not by an ice age but through the amelioration of the climate. Softer conditions allowed a competitor into Europe.

Neanderthal man is believed to have become extinct about 35,000 years ago, although a few anthropologists have maintained that Neanderthal skeletal formations have persisted through a genetic mixing of human subtypes and that people with Neanderthal characteristics are walking the Earth today, particularly in Scandinavia. Be that as it may, Neanderthal man largely died out, to be replaced in Europe by our own subspecies of humanity.

In 1886 the French geologist Louis Lartet began excavations at Cro-Magnon, near Les Eyzies-de-Tayac in the Dordogne, that led to the discovery of human remains indistinguishable from those of a modern European. Cro-Magnon man lived about 35,000 years ago. Later discoveries in Poland and Hungary set the appearance of similar men also at about 35,000 BP, close to the time when Neanderthal man disappeared.

There is no evidence that Cro-Magnon exterminated Neanderthal man, as has been suggested, but the appearance of the one at about the time the other disappeared is suggestive. If Cro-Magnon man had come from a warmer climate to the south, his penetration into Europe would need to have coincided with an amelioration of the climate, since under ice-age conditions and with only primitive technology he would not have been able to compete with the hardy Neanderthal man. Indeed, a warm period would have worked to the detriment of Neanderthal man in two ways: besides generating competitive pressure from the south, it would also have produced serious disorganization in Neanderthal man's own social structure.

At first sight it might seem strange that an amelioration of physical conditions could lead to weakening, but that is just what happened to Hannibal, the Carthaginian general, when he was unwise enough, in 216 BC, to winter his army in the soft Capua to the north of Naples. It is also what happened to Cro-Magnon's own descendants when warm conditions returned to Europe at 10,000 BP. The high Magdalenian culture of 15,000 BP disappeared with the last ice age. Soft living does not appear to be good for humans, as one may easily verify from observing the modern scene.

A strange thing about humans is their capacity for blind rage. Rage is presumably an emotion induced by survival instinct, but the surprising thing about it is that we do not deploy it against other animals. If we encounter a dangerous wild animal – a poisonous snake or a wild cat – we do not fly into a temper. If we are unarmed, we show fear and attempt to back away; if we are suitably armed, we attack, but in a rational manner not in a rage. We reserve rage for our own species. It is hard to see any survival value in attacking one's own, but if we take account of the long competition which must have existed between our own sub-species and others like Neanderthal man – indeed others still more remote from us than Neanderthal man – human rage becomes more comprehensible.

In our everyday language and behaviour there are many reminders of those early struggles. We are perpetually using the words 'us and them'. 'Our' side is perpetually trying to do down the 'other' side. In games we artificially create other subspecies we can attack. The opposition of 'us' and 'them' is the touchstone of the two-party system of 'democratic' politics. Although there are no very serious consequences to many of these modern psychological representations of the 'us and them' emotion, it is as well to remember that the original aim was not to beat the other subspecies in a game but to exterminate it.

The readiness with which humans allow themselves to be regimented has permitted large armies to be formed, which, taken together with the 'us and them' blind rage, has led to destructive clashes within our subspecies itself. The First World War is an example in which Europe divided itself into two imaginary subspecies. And there is a similar extermination battle now in Northern Ireland. The idea that there is a religious basis for this clash is illusory, for not even the Pope has been able to control it. The clash is much more primitive than the Christian religion, much older in its emotional origin. The conflict in Ireland is unlikely to stop until a greater primitive fear is imposed from outside the community, or until the combatants become exhausted.

The descendants of Cro-Magnon man left a great deal of information about themselves. The information is so complex, however, that it has not yet been satisfactorily interpreted. From Germany to the Ural mountains their symbolic carvings and sculptures have been recovered, while in the limestone caves of central France, through the Pyrenees and into Spain, their magnificent cave paintings survive.

The first discovery of cave paintings was made in 1868 – appropriately enough by a modern hunter – at Altamira in northern Spain. The Altamira cave paintings, dating from about 15,000 BP, are still the outstanding examples of Magdalenian polychrome art. Although the cave is nearly 300 metres long, most of the paintings are in a chamber 18 by 9 metres, which has an average height of 2 metres. The roof of the chamber is covered by paintings of bison in red, black and violet, with wild boars, a hind and other figures in less detailed style. These extremely literal animal forms are accompanied by symbolic human figures and other less clear markings.

There has been much argument about the meaning of these paintings. An early theory, still widely quoted, holds that the caves were the scenes of magical hunting rites, that they were the temples and cathedrals of the prehistoric world. The paintings themselves are thought to be the sacred texts in picture form, like the ikons of Byzantium. This theory is psychologically sound. The attachment of mystic significance to symbols, words and objects is still widespread today. I have already mentioned 'the sceptre' and I could add many other examples, such as the font, the altar, the crown of a monarch or the city mayor's chain of office.

Against the magic hunting interpretation of prehistoric art, some archaeologists have argued that the animals portrayed most frequently are not the animals that were eaten most often. The answer to that objection is simple: the foods we eat every day are rarely the ones we prize most highly. At all events, it is generally agreed that the paintings were not art for art's sake.

It is widely believed that, in causing the pyramids of Egypt to be constructed, the Pharaohs were motivated by the desire to immortalize their memory. Other examples of the pursuit of immortality are widespread in written history. Even our own century is not without them; when a year or so ago a NASA space vehicle first left the solar system, a plaque was included in the package inscribed with a symbolic representation of the Earth and it inhabitants. This was done on the million – to – one chance that the vhicle might be intercepted by some other intelligence somewhere in the Galaxy, thereby immortalizing the memory of the human species.

Something of the same idea may have motivated the descendants of Cro-Magnon man. Each cave may have been a visual display of their world. It could not have been an encyclopedic display, of course, but a display of the things which seemed most relevant. The artists made no attempt to paint natural scenery as a background to the animal pictures, for the obvious reason that natural scenery would have seemed invariable. The animals on the other hand were changing in their kinds and numbers from decade to decade. The people themselves are not shown in a natural way, probably because the actual appearance of people would also have seemed invariable. Other aspects of the lives of people that had an unchanging quality would also have been represented symbolically, since their significance would have been thought obvious to future generations. Whatever the purpose of the paintings actually was, the fact is that Cro-Magnon man did indeed immortalize himself. And, if archaeologists should learn to read the meaning of the many symbols, they may well approximate to written history.

There can be little doubt that the great technical ingenuity man has displayed over the last 5000 years is a heritage from past ice ages. To have survived the fearsome conditions of those years, our distant ancestors had to be both extremely hardy and very clever. There can have been no niche for the weak or the foolish. When the early decline that followed the end of the last ice age was reversed after about 8000 BP, the improvement, at first gradual, was eventually able to swell to a mighty flood. It is a mistake to think that there is anything to choose between the ability of a modern Newton and that of the geniuses of prehistoric times. The main difference lies in the veneer of modern education. If you wish to get an idea of the real driving quality of our ice-age ancestors, perhaps the best way is to read the life of George Stephenson, for Stephenson showed the full measure of the untutored brilliance we inherit from the past.

This astonishing tale of man's success has a still more astonishing denouement. It has turned out that man has a quite extraordinary weakness. He is very suggestible; if he is told he is stupid, he thinks he is stupid and behaves accordingly; if he is told he needs cossetting, he thinks he needs it; he has allowed himself to be enslaved throughout history; and in Victorian times his womenfolk were told they were weak and they believed it – women whose distant grandmothers survived in 30°, 40° or even 50° of frost, while they and their men disputed the possession of caves with bears and mountain lions.

The strange human animal is susceptible to endless regulation. Scarcely a day passes but regulators talk about protecting the people, even about protecting the people from themselves. But people whose ancestors fought off the wildest animals of the Earth with puny sticks do not need protection – except from regulators. Animals of this species, almost any one of which is capable of extraordinary feats if roused, sink into a torpor when drably regulated. Their productivity falls ever lower, until in the end they are in danger of joining the giant sloth, an animal which manages to shift itself, when hard pressed, at a speed of 2 or 3 feet in an hour. Fortunately, the next ice age will change men back to what they used to be.

3 The Ice-Age Controversy

Ice ages have always provoked argument. In the first half of the nineteenth century even the suggestion that there had once been an ice age was greeted with derision. Since the scientific world eventually came to accept the former existence of ice ages, the controversy has been concerned more recently with the question of why the climate of the Earth fluctuates so drastically from one epoch to another.

Local evidence of the ice age of 20,000 BP is so widespread throughout northern temperate latitudes that it is impossible to say who in modern times first conceived the astonishing idea that a great ice sheet once covered his or her home town or village. The thought must, I think, have occurred to many peasants and small farmers long before it came to be expressed in a literate form.

I myself have seen a host of examples of glacial action throughout the Scottish Highlands and the English Lake District. Rivers by themselves cut deep V-shaped valleys in a mountainous topography, as illustrated in Figure 10. Subsequent glaciation in Figure 10B converts the V-shaped valleys into U-shaped valleys, which remain when the glaciers eventually melt away (Figure 10C). So it is U-shaped valleys that one looks for in seeking evidence of past ice ages.

Possibly the finest example I have seen is the fjord of Loch Leven in Scotland. The best viewpoint requires one to climb for about a mile up the hillside to the north-east of the town of Kinlochleven. From there, one can see that, besides being perfectly U-shaped, the valley also has symmetrically placed pimples of about the same height as each other on either side. The one seen to the left is the Pap of Glencoe, well known to thousands of passing motorists; the other is Mam na Gualainn on the north side of the loch. No one could pass off this matching as

41

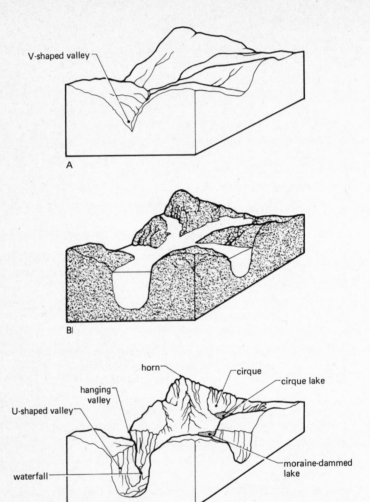

Figure 10: *Glacial erosion of a mountainous region. Erosion by running water makes V-shaped valleys (A). Glaciation (B) produces the characteristic ice-eroded forms (C).*

chance. The little peak on the north side of the valley must have been formed in association with the Pap, and it must have been formed by something which rose to the heights of the little peaks themselves, about 2500 feet (about 800 metres) above the loch. The agent could hardly have been water. It could only have been ice – a glacier about 800 metres thick, which once flowed around the peaks like a river flowing around a stone projecting from its bed. The evidence of the scraping and scratching of ice on the summit is unmistakable in the clear-cut terrace of the Pap.

It was, however, evidence of glaciation in the Jura Mountains that sparked off the first ice-age controversy. The mountains run about 300 kilometres along the Franco-Swiss border, from south-west to north-east. The rocks of the Jura were laid down during the geological 'Jurassic' period (190 to 135 million BP) that takes its name from the region. The mountains are far older than the main Alpine chain and therefore much smaller, having been greatly eroded over past ages. They are extensively wooded because of their comparatively low elevation – hence their name, for *jura* is an old Gaulish word meaning 'forest'. Jurassic rocks are mainly limestones and they give sweet, fertile soils.

It must have been obvious to many people in the Jura towns of Neuchâtel and Solothurn that the granite boulders to be found scattered over their valleys and hillsides were quite different from the surrounding limestone rocks. Indeed it must have been a matter of common knowledge in the district that those granite boulders had the same look about them as boulders from valleys 75 kilometres away to the south-east, boulders native to the much higher mountains of the Bernese Oberland. The idea that boulders had somehow been transported from the region of the Oberland – a concept that was to torment a later generation of savants – must have been commonplace to the ordinary folk of Neuchâtel and Solothurn, and of course to the inhabitants of many other towns and villages of the Jura.

The idea that the granite boulders found in the Jura had been transported by Alpine glaciers that once spread as far as 100 kilometres to the north seems to have been published first in 1787, by a Swiss pastor, Bernard Friedrich Kuhn. In 1795, James Hutton, the 'father of geology', made the long journey from Edinburgh to the Jura to examine the evidence. He was in no doubt about the meaning of the granite boulders he saw around Neuchâtel, and he stated unequivocally that huge glaciers must once have extended outwards from the Alps. This pronouncement should have settled the matter there and then, but in his own time Hutton's work went largely unrecognized; his opinions seemed much too extreme, too wild to be taken seriously, for they implied that the Earth must be enormously older than his blinkered generation was prepared to believe. So Hutton's confirmation of Kuhn's suggestion was ignored.

In the lower Alpine valleys that are now free of ice, there are many relics of a former ice age. There are expanses of rock 'pavements', smoothed and scratched by the abrasive action of debris carried by moving ice. There are smoothed boulders and there are strips of hummocky ground, known as moraines, similar to those found in the higher, still glaciated, parts of the valleys. These relics show clearly that the glaciers must once have extended far down the valleys. It was inevitable that, with generally increasing literacy, observant dwellers in these valleys would come to join in the great controversy. In 1815 or thereabouts Jean-Pierre Perraudin, a chamois hunter from the Val de Bagnes, wrote as follows:

> Having long observed scratches on hard rocks which do not weather, I at last decided . . . that the marks had been made by the weight of glaciers. Since I find such traces at least as far as Champsec I believe that in the past glaciers filled the whole Val de Bagnes, and I am prepared to prove my belief to the incredulous by comparing the scratches with those uncovered by glaciers at the present time.

Perraudin's arguments were greeted with scepticism, but were eventually accepted by Ignaz Venetz, a civil engineer. Venetz had opportunities to collect evidence over a wider area than one particular valley. Although he began by accepting the statements of Hutton and Kuhn, in 1829 he suggested that the Alpine glaciers had extended even further to the north, across the Jura Mountains into the European plain. By then, Jens Esmark, working independently in Norway, had concluded that the Norwegian glaciers had been far greater than their modern, shrunken residues.

In 1834 Jean de Charpentier read a paper in Lucerne at a meeting of the Swiss Natural History Society in support of the position taken by Ignaz Venetz, giving as evidence extensive field observations of his own. Apparently endowed with a lively sense of humour, de Charpentier took care to emphasize his earlier scepticism, poking fun not only at himself but also, by implication, at his doubting audience. He included the following anecdote:

> Walking through the Hausli valley on the road to Brunig I met a woodcutter from Meiringen. As I was examining a large boulder of Grimsel granite lying by the roadside, the fellow said, 'There are many stones like that around here, but they come from far away, from the Grimsel, because they are made of Geisberger, and the mountains here are not made of it.'
>
> When I asked him how he thought that these stones had reached their location, he answered, 'The Grimsel glacier brought them and dropped them on both sides of the valley. The glacier once reached as far as Berne.'

But, in spite of the woodcutter, de Charpentier did not carry his audience. One member of it in particular, a man of very forceful personality called Louis Agassiz, refused to accept his conclusions. Over the next two years, Venetz and de Charpentier took the trouble to accompany Agassiz into the field, where they showed him their evidence and explained it in great detail. What Agassiz saw opened his eyes and, like Paul after his vision on the road to Damascus, he devoted the rest of his life to converting the world to his new faith. In his enthusiasm for the belief that there had indeed been an ice age, Agassiz proclaimed that its glaciers had reached even to the Earth's equator.

Visiting Britain in 1840, Agassiz succeeded in converting the Reverend William Buckland, professor of geology at Oxford. Buckland, who had originally rejected the glacial theory, was persuaded to visit the Lake District, where he found abundant evidence of glacial moraines, particularly near Eden Hall, to the east of Penrith. Agassiz and Buckland travelled to Scotland together, and there – even to the still partly unconvinced Buckland – the evidence was overwhelming. On 7 October 1840, the *Scotsman* carried an article entitled 'Discovery of the Former

Existence of Glaciers in Scotland, especially in the Highlands, by Professor Agassiz'. This article, deeply shocking to many, set the ice-age controversy roaring like an Atlantic gale over the British Isles. It is hard to believe that, with so much decisive evidence readily to hand in as small an area as Britain, the argument could have persisted for more than a year or two. But persist it did, until it was at last settled by two great papers, one published in 1862 by T. F. Jamieson, the other in 1863 by Archibald Geikie.

Agassiz visited the United States in 1846 and settled there to take up a professorship at Harvard University. The ice-age controversy was one of the first to be conducted in the full glare of what we now call media publicity, and perhaps for this reason the story was distorted as it passed into popular history. Louis Agassiz did not invent the ice-age theory. All the important pieces of the theory were already in place before he ever came on the scene. His principal contribution was to stir up great general interest in the theory in Britain and among geologists in the United States. It has sometimes been said that, without Agassiz's powerful advocacy, the theory would not have come to be accepted, but it was clearly gaining momentum anyway and would have triumphed in the end. I find it sad that, while the usually comprehensive modern edition of the *Encylopaedia Britannica* contains a full text biography of Agassiz, there is no mention even among the shorter biographies of the real pioneers: Kuhn, Perraudin, Esmark, Ignaz Venetz, de Charpentier, or the woodcutter from Meiringen.

Once geologists in the United States and northern Europe had accepted the ice-age theory, they became fascinated by the study of erratic boulders. Enormous numbers of boulders of distinctive rock were found at remarkably large distances from their places of origin. It may seem surprising that stones can be different enough from one another to permit their sources to be closely identified, and indeed this is not always the case, but some stones such as Shap granite are unique to particular surface outcrops. Shap granite pushed its way upwards some 400 million years ago and now only outcrops the surface over an area of about 3 square miles between Kendal and Shap in the English Lake District. Once one has seen this pinkish stone with the enormous crystals embedded in it, one could hardly fail to recognize it, even as far from its outcrop as the Yorkshire or Durham coasts. Another highly distinctive erratic, a beautiful greenish-blue stone, comes from an outcrop near the Norwegian seaport of Larvik. The region around Oslo has produced other notable distinctive rocks. One boulder of a red, rather coarse, granitic type of porphyry was used for the sarcophagus of Charles XIV of Sweden. The boulder was found at Eldfal in Sweden, where it was carved into shape. The work took twelve years and, when it was at last finished, the heavy coffin had somehow to be transported all the way to Stockholm. In the end it was loaded on to a sledge drawn in winter by hundreds of peasants. They are said to have been encouraged in their task by a fiddler sitting on top of the load.

The erratics of the Jura Mountains that sparked off the whole ice-age controversy were thought to have been carried by glaciers that had spread north from the high Alps, through distances of about 100 kilometres. When erratics were later found to have travelled more than 100 kilometres from their sources, geologists merely assumed that the glaciers had spread even further than they had at first thought. Even when erratic boulders from Norway were found on the English coast, scientists accepted without question that glaciers had spread across the North Sea.

Some smoothed, striated boulders similar to that shown in Figure 11 must indeed have been transported by glacier movement. I can well imagine a boulder detached by frost action falling on to the surface of a glacier and being carried downhill without its sharp edges becoming smoothed. But I do not believe that far-travelling erratics, such as that shown in Figure 12, with sharp edges and many facets, could possibly have been pushed or dragged at the bottom of a huge glacier over great distances without losing their angularities. Glacier till consists of stones of varying sizes embedded in a considerable quantity of fine-grained material. When a glacier melts, it deposits its original stony contents from various depths in one mixed heap (which explains the generally unsorted nature of glacier-deposited material) and many of the stones of such a till have been generally smoothed and scratched by their passage in the glacier – as in Figure 11, not as in Figure 12.

Towards the end of the nineteenth century, D. Mackintosh

wrote many splendidly voluminous papers on British erratics in which he frequently mentioned the angularity of boulders he had found (for which he sometimes gave compass bearings laid off from the nearest pub). He also plotted the tracks of boulders deposited in the Midlands. To his surprise, he found that the lines of boulders from mid-Wales ran directly across those of stones from south-west Scotland. If they had been carried by glaciers, this would have meant that glaciers coming from different directions had ridden across each other, apparently without interference.

Figure 13 demonstrates another difficulty of the glacier theory. It implies that glaciers flowed in all directions across Ailsa Craig, a small intrusion of granite in the Firth of Clyde.

Figure 11: *A glacier-smoothed striated boulder, from J. Geikie's classic* The Great Ice Age *(1877).*

Although Ailsa Craig rises to 1115 feet (340 metres) above the sea, its area is a mere three-eighths of a square mile (1 square kilometre). Obviously Ailsa Craig could not have attracted enough precipitation (under the dry conditions of an ice age) to supply ice for glaciers which spread out radially for more than 50 miles (80 kilometres). The evidence of Figure 13 – that erratics have spread out radially from a small eminence – is supported by a dozen or more similar diagrams of other areas.

Mackintosh himself believed that the erratics were transported by ice-rafting, a process Charles Lyell had suggested many years earlier. They believed that the stones were carried in or on a lump of ice that floated in water. But ice-rafting could never have carried erratics uphill, and Shap granite boulders have somehow travelled 1000 feet (300 metres) up over the Stainmore Pass into Durham and Yorkshire, while in Scotland blocks of Torridon sandstone have been found 1500 feet (460 metres) higher than their source. In America, erratics have been discovered up to 1500 metres higher than their sources in the

Figure 12: *An erratic boulder, from J. Geikie's* The Great Ice Age. *Note how different it is from the smoothed stone of Figure 11.*

mountains of New Hampshire. The mystery of how they were transported is not solved by the theory of ice-rafting, but at least Mackintosh saw there was a mystery to be solved – a perception which probably did not endear him to his contemporaries, who were overwhelmed with enthusiasm for the glacier-flow idea.

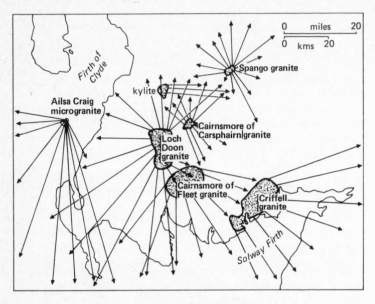

Figure 13: *The distribution of erratics carried from source deposits in south-west Scotland.*

When the Norwegian boulders were discovered along the coast from Durham to Norfolk, they gave rise to a theory that vast outlet glaciers from the Norwegian mountains had extended for 1000 kilometres across the North Sea. Although this view is still widely held by geologists and glaciologists, it appears to me to contradict what we know about the physical properties of ice, as I hope to demonstrate in the rest of this chapter.

When water vapour evaporated from the ocean surface condenses in the atmosphere into water and ice crystals, it begins a train of events of astonishing complexity. Temperature is without losing their angularities. Glacier still consists of stones of atmosphere that determine various kinds of ice crystal. The details of this relationship are set out in Table 1.

Table 1: **Relation of ice-crystal shape to temperature of formation**

temperature (°C)	form
0 to − 3	thin hexagonal plates
− 3 to − 5	needles
− 5 to − 8	hollow, prismatic columns
− 8 to −12	hexagonal plates
−12 to −16	dendritic crystals
−16 to −25	hexagonal plates
−25 to −50	hollow prisms

Underlying the general trend towards hexagonal forms is the basic internal structure of ice. The water molecules (each with two hydrogen atoms and one oxygen: H_2O) arrange themselves into primary groups of six, with the oxygen atoms of each group linked into a hexagon. In the next stage, the hexagons link together to form a sheet, and in the third stage the many sheets join up to one another like the pages of a book. The oxygen atoms in parallel hexagons then become interlinked, causing the 'pages' of the 'book' to stick together. This last step, this sticking together of the pages, is weaker, however, than the joining of the hexagons in the individual sheets, and it is the comparative weakness with which the pages are stuck together that gives ice the interesting mechanical properties that affect the behaviour of mountain glaciers and ice sheets. None of the linkages of ice molecules is at all comparable in strength to the linkages of the atoms in a piece of metal or rock, which is why ice melts at a much lower temperature than do rocks or metals.

The linkages of the atoms inside a solid can be thought of as springs that can be compressed or extended or tilted at various angles. They hold the solid together against external forces (for example against the weight of the solid) – external forces that would otherwise immediately pull the solid apart. The springs between the atoms are always in some degree of compression (pressure) or extension (tension) or of varying tilt (shear).

51

There is a good way to think about the three different kinds of force inside solids. Imagine yourself putting a smooth plane through a solid. If you then find an inward force on your plane, acting on both sides of it, there was pressure in the solid. This would be true of a horizontal plane put through a lump of lead lying on the ground. If, on the other hand, the two pieces into which your plane divides the solid immediately fall apart from each other, then there was tension inside the solid. This would be true of a lump of lead suspended from the hook of a crane. Tension and pressure are really the same kind of force – they are negative and positive values of the same force – but shear is quite different. If, when you put your smooth plane through a solid, the two resulting pieces slide against each other, then there was shear inside the solid.

Most of us have had the unpleasant experience of taking a spill from a moving bicycle. The injuries one suffers are visible examples of the difference between pressure and shear. The bump one experiences on hitting the ground produces a bruise caused by a pressure force between the body and the ground. As well as the bruise, there is usually a scoring of the skin and perhaps some severe scratches, caused by the sliding which produced a shear force along the skin.

It is possible to put a plane through a body in an infinity of ways, and in general each plane gives a different result, so that, when one speaks of pressure or tension or shear inside the body, it is necessary to specify the particular cutting plane one has in mind. Consider a uniform cube of solid material resting on one of its faces on the ground. It is easy to see that neither a horizontal plane nor a vertical one put through the cube will cause any sliding. But a smooth plane put through the cube at an oblique angle – say a plane with a slope of 10 degrees to the horizontal – would cause sliding. So an oblique plane causes shear. It is impossible to describe the forces inside a body except by reference to cutting planes.

All bodies collapse in one way or another when the forces inside them become large enough, and they all react to pressure or tension in similar ways: put a body in an open press and increase the pressure steadily, and sooner or later the material of the body will be squeezed like toothpaste out from the sides of the press. All solids crack open under a sufficiently large tension, but all bodies do not behave in the same way with respect to shear. As the shear across a particular plane inside a metal is increased, the linkages of atoms across the plane tend to break, and for a sufficiently large shear a crack develops along the plane. There will be a particular plane along which the shear is greatest, and it is along this plane of greatest shear that the crack actually develops. Because no metal is entirely uniform, the shear forces are not uniform either, and so the cracking begins locally, where the shear is largest or the metal weakest. Once a crack has opened, the small broken region can no longer help to prevent sliding, with the result that other regions have to sustain an increased shear. This causes some other place to crack, weakening the metal still more; a 'hairline' crack develops. The cracking proceeds step by step, until at a sudden moment the whole thing goes, an aircraft engine tears away or a bridge collapses from 'metal fatigue'.

The situation with ice is different, however. When linkages between neighbouring atoms are broken by shear, the linkages soon re-form again, with the atoms systematically changing partners in the brief moments when the linkages are broken. A kind of slow slide known as 'plastic creep' ensues, which is why glaciers flow slowly down the flanks of mountains.

Ice is exceptional among plastic substances in the important respect that its rate of creep increases rapidly with the strength of the shear. If one were to take cubes of graduated sizes of some plastic material, placing them each with a flat face resting on a horizontal surface, the small cubes would hardly change at all, but the large ones would slump because of the plastic creep caused by their weight. Think of a cube of treacle; it is clear that the treacle would flow in a smooth bead spreading out and flattening itself. If you were to measure the time it took for different-sized cubes of the same material, you would find that the larger ones collapsed into a bead of, say, half the height of the original cube more quickly than did the small ones. For a normal plastic substance, doubling the cube size halves the collapse time: but, for ice, doubling the cube size shortens the collapse time by eight, according to the experimental law put forward by J. W. Glen in 1955. Ice weakens rapidly as the shear increases.

53

This property is responsible for a great deal of the dramatic behaviour of mountain glaciers.

Suppose a mountain glacier to be of a uniform thickness, and suppose also that it rests on a rough bed of uniform downward slope. The forces arising from the weight of the ice are obviously greatest at the bottom of the glacier. This means that, if we were to take a plane through the ice parallel to the bedrock, the shear along the plane would be greater if the plane were near the bottom of the ice than if it were near the surface. According to Glen's law, the plastic creep rate would therefore be very much greater towards the bottom of the ice than towards the top. Measurements on actual mountain glaciers support this.

The Athabasca Glacier, which is splendidly visible from the road which links Banff to Jasper in the Canadian Rockies, is about 200 metres thick. Owing to the slope angle of approximately 8° at which this glacier descends from the Columbia Icefield, the uppermost 100 metres of the ice is moving downwards at a speed of about 30 metres per year. The slip against the rough bedrock is much slower than this – no more than a few metres per year – as can be seen in Figure 14. The slide happens because of plastic creep occurring in the lower 100 metres of ice – indeed, with most of the creep in the bottom 50 metres, just as one would expect from Glen's law. Mountain glaciers move downhill mostly through plastic flow in the ice itself, not by scraping bodily down the bedrock.

The South Cascade Glacier, also on the Pacific north-west coast of North America, is about 100 metres thick and it also flows down a slope of about 8°. The downward speed at the surface averages only about 8 metres per year, however – a quarter of the downward rate of flow of the Athabasca Glacier. This is because the lesser thickness of ice generates less shear near the base of the glacier. For glaciers of the same slope (and with ice of the same temperature), the downflow increases rapidly with the thickness of the ice. Calculation according to Glen's law shows that increasing the thickness from the 200 metres of the Athabasca Glacier to the 2 kilometres of the world's largest glaciers would increase the flow rate ten-thousandfold – from 30 metres per year to nearly 1000 metres per day. The flow would become a veritable cascade of ice. It is clear that mountain glaciers cannot, except in a few most unusual cases, flow at anything like this speed, because the supply of snow on the upper slopes of a mountain is hardly ever sufficient to provide the enormous amount of ice that would be needed to feed such a cascade. The thickness of a mountain glacier must necessarily depend on the source of its ice. For this reason, under present-day conditions, the thickness of most steep mountain glaciers is restricted to not more than a few hundred metres.

The shear with respect to a plane parallel to the slope is less for smaller angles of slope than for larger ones and, because of the sensitivity of the ice flow to the strength of the shear, downward speeds of flow fall off sharply with a reduction in the slope angle.

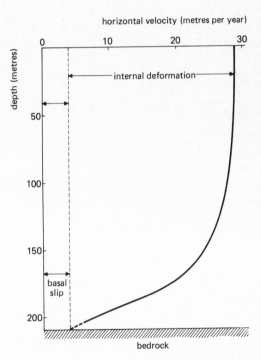

Figure 14: *Speed of motion of ice at varying depths in the Athabasca glacier, Canada. Essentially the whole of the motion is generated by the shearing force in the bottom 50 metres. (After J. C. Savage and W. S. B. Patterson*, J. geophys. res., *vol. 68, 1963, p. 4521.)*

Thus, for a slope angle of 0.8° instead of 8° the flow rate is reduced a thousandfold, from 1000 metres per day for a glacier 2 kilometres thick to about 1 metre per day – as, for example, in the case of the great polar glaciers, such as those which cut through the Transantarctic mountains in their descent to the Ross Iceshelf. One of them, the Beardmore Glacier, is about 200 kilometres long and 23 kilometres wide. Its flow rate is about 1 metre per day. If a glacier is very thick, its slope angle must be small. The majestic Byrd Glacier is an extreme example of a very thick glacier on a small slope: it is rather more than 2 kilometres thick and its slope is less than 0.5°.

A process exists today in regions of low precipitation that can limit glacier thicknesses to about 2 kilometres, and the same process may have operated during the ice ages. At a certain thickness, depending on the surface temperature of the ice, geothermal heat – the heat that escapes continually from the interior of the Earth – becomes trapped at the bottom of a glacier. For a typical geothermal heat flow (70 milliwatts per square metre), trapping occurs for a surface temperature of −10°C when the glacier thickness exceeds about 400 metres; for a surface temperature of −30°C, trapping occurs at a thickness of 1.3 kilometres; and, for a surface temperature of −50°C, at a thickness of 2.1 kilometres.

Geothermal heat raises the temperature of the lower regions of a glacier and, if the glacier is thick enough, the temperature at the bottom rises sufficiently for ice to begin melting. Thereafter, the geothermal heat melts the bottom ice, so becoming 'trapped'. The resulting water tends to be squeezed out of the sides of the glacier or, in the case of a glacier enclosed between rock walls (as in a fjord), the water tends to be squeezed along the length of the glacier. Each year, the geothermal heat is sufficient to melt a layer of bottom ice about 1 centimetre thick, equivalent in quantity to an annual snowfall of about 10 centimetres (snow being very light fluffy stuff, it takes about 10 centimetres of new snow to make 1 centimetre of ice). If the rate at which snow accumulates at the surface exceeds this amount, the glacier continues to grow in depth but, if the accumulation of snow is less than 10 centimetres each year, as it may be under very dry conditions, the depth becomes limited when geothermal heat becomes trapped.

Melting produced by geothermal heat occurs nowadays at the bottom of the Byrd Glacier.

We have considered the way a large cube of material like ice slumps if it is placed on a horizontal plane. A huge cube of ice slumps into a bead that spreads itself, but eventually the spreading stops when the forces acting on the ice come into a close equilibrium. The ice then forms a more or less stable mound with a radius determined by the amount of ice. The mathematical problem of determining the radius of such a stable mound was solved about a quarter of a century ago by J. F. Nye.[1] Take the height of the ice at the centre of the mound – say 'h' metres – square the height, divide by 22 metres, and you have the required radius (in metres) of the ice mound. Nye's solution for calculating the radius is easily applied to the Norwegian glaciers that are supposed to have carried boulders across the North Sea. For outlet glaciers with a height at the Norwegian coast of 2 kilometres, $h = 2000$ and $h^2 = 4$ million square metres. Dividing by 22 metres gives about 182,000 metres, i.e. 182 kilometres, much less than the required 1000 kilometres. The result corresponds very well, however, with the lengths of the great Norwegian fjords and, while this is not clear-cut proof that the result is correct, it provides strong support for Nye's theory. Figure 15 compares a calculated ice-mound profile (for h close to 2600 metres) with the measured profile of an actual mound in north Greenland. The correspondence is close enough to exclude the possibility of gross error in the theoretical method.

The above remarks assume that the ice was hard-frozen, not a mere pile of mush, but this condition must have been well satisfied during an ice age, with the temperature in Norway then much like it is in Greenland today.

Ice can, of course, flow any distance, as long as it does so on a downward slope, but across the North Sea we are not concerned with a downward slope. The slope actually rises about 200 metres towards the British coast. The problem cannot be resolved by supposing h to have been significantly larger than 2000 metres. Apart from the limiting effect of geothermal heat, there is plenty of evidence from Scandinavian land-forms that this

[1] J. F. Nye, *Nature*, vol. 169, 529, 1952.

estimate for the outlet glaciers as they reached flat ground is correct. The weight of the ice-age glaciers depressed the whole Scandinavian area to the extent that it is still rebounding today. This effect has been carefully studied and is consistent with glaciers about 2000 metres thick. If they had been much thicker, the depression of the land would have been much greater than it actually was and the present-day rate of rebound would have been faster.

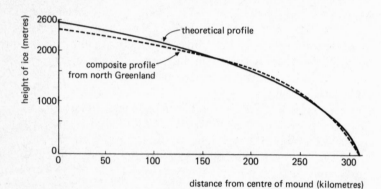

distance from centre of mound (kilometres)

Figure 15: *Profiles of an ice mound. The theoretical profile obtained by J. F. Nye compares closely with the composite profile obtained from measurements in north Greenland.*

It is also impossible to believe that, in Britain, Lake District erratics were carried to the Midlands by glaciers. Mackintosh, who wrote in 1879 that the number of erratic boulders near Wolverhampton was so large that the people used them as paving stones, had his own reasons for disbelieving the glacier theory, but many people have continued to accept it.

It has been estimated that the maximum thickness of the Lake District's glaciers was 2000 feet (600 metres). Applying Nye's formula, it is clear that they could only have extended about 10 miles (16 kilometres). The fact that the largest lake, Windermere, is just about 10 miles (16 kilometres) long suggests that these calculations are correct. Some authors have offered even lower estimates for the depth of the glaciers. For example, Millward and Robinson wrote:

> Many of the major features seen by those visiting Great Langdale date from a time when a great glacier occupied the whole valley during the last glaciation. There is evidence to suggest that at its maximum stage of development the upper surface of the glacier lay at a height of about 1400 feet.[2]

Furthermore, geological investigations of the region of Muncaster in west Cumbria have shown that glaciers from the Scafell region extended only about 10 miles (16 kilometres) from Wasdale Head. The *unglaciated* fellside of Black Combe a little further to the south is clearly outside the zone of the Scafell ice. Such unglaciated regions prove that the Lake District outlet glaciers could not have been thicker than 2000 feet (600 metres) and that the erratic boulders that Mackintosh found in such profusion near Wolverhampton must have been transported by something other than glacier flow.

The lengths of the longest glacial lakes in a previously ice-covered area tell us the depths of the former outlet glaciers immediately: 10 miles (16 kilometres) in the Lake District for a depth – according to Nye's formula – of about 600 metres; a length generally of 15 miles (25 kilometres) in the Highlands of Scotland (Loch Arkaig, Loch Ericht, Loch Maree) for an ice depth of about 730 metres; a length of about 35 miles (55 kilometres) for the longest finger-lakes of New York State giving a depth of ice of about 1100 metres; a length in Norway and Sweden of about 80 miles (130 kilometres) for a depth of about 1700 metres; and a typical length for lakes of glacial origin in Canada of 200 miles (320 kilometres) for a depth of ice of about 2700 metres. Although these estimates should be considered approximate, they do indicate unequivocally that the ice sheets of the last ice age could not have been caused by the flows of spreading glaciers.

Bernard Kuhn and James Hutton were probably correct in stating that the Jura erratics had been carried by glaciers from the Swiss Alps. A map of Switzerland shows that the major Swiss

[2] *The Lake District*, Methuen, 1970.

lakes are all about 75 kilometres long, implying that the outlet glaciers had a depth of about 1200 metres. They must have reached out from the base of the valleys by approximately 75 kilometres – just about the distance travelled by the Jura erratics.

Jean-Pierre Perraudin, the chamois hunter, was certainly correct in his statement that a glacier once filled the Val de Bagnes, but Ignaz Venetz and Jean de Charpentier went too far. The Alpine glaciers could not have flowed out beyond the Jura Mountains. Louis Agassiz was right in his article of 1840 in the *Scotsman*, but his grand concept of glaciers swooping from the mountains and covering Europe and the United States was quite wrong.

Just as the erratics could not have been carried by glaciers, the ice that once filled the Irish Sea and reached to the English Midlands, the ice which once covered the European plain as far south as Brandenburg, and the ice which once covered the American Middle West south of Chicago, cannot have come from spreading glaciers. It must have been deposited in some other way.

4 Astronomical Theories of the Cause of Ice Ages

The mistaken belief that ice ages were caused by an enormous expansion of mountain glaciers led to a discussion of the factors that determine the extent to which a mountain glacier extends down into its associated valley. The two important things are the feed of snowfall at the top of a glacier, and the melting and evaporation near the bottom. Increasing the feed or reducing the melt (or both) would extend a glacier, while decreasing the feed or increasing the melt (or both) would reduce it. Since melting is confined to the summer months, it appears natural to think of the melt as being controlled by the summer sun. And, with much of the Alpine snowfall occurring in winter, the idea soon developed that the size of a glacier was determined by a battle between winter snowfall and the summer sun. The inevitable question was: had there been more snowfall during the ice ages or had the summer sun been weaker? Opinions differed and the resulting arguments were strongly contested. Eventually, the majority view emerged that a weakening of the summer sun must have been the principal cause of the advance of glaciers and hence of the ice ages themselves.

Figure 16 explains why we experience summer and winter. The Earth goes once round the Sun each year, moving along an orbit that is nearly a circle. Each day, meanwhile, the Earth makes a complete turn around its polar axis, which is tilted to the plane of the orbit (the so-called ecliptic) in the manner shown schematically in Figure 16. For the northern hemisphere, the axis of rotation leans most towards the Sun on 22 June, and, for the southern hemisphere, most towards the Sun on 22 December. These dates correspond respectively to midsummer for the two hemispheres. Thus 22 June is midsummer for the northern

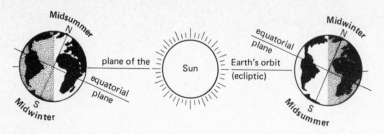

Figure 16: *The variation of solar radiation during the year. As the Earth moves in its orbit round the Sun, the angle between the polar axis of rotation and the direction of the Sun varies. For each hemisphere, the angle is least at midsummer and greatest at midwinter.*

hemisphere and midwinter for the southern hemisphere, with the opposite applying on 22 December.

The Earth's orbit is slightly elliptical, which means that at some times in the year the Earth is a little nearer to the Sun than at other times. It is nearest in early January, and farthest away in early July. The effect of these variations is to make northern hemisphere winters a little warmer, and the summers a little cooler, than would be the case if the Earth followed a strictly circular path around the Sun. Radiation from the Sun is increased about 3 per cent in the northern winter and decreased about 3 per cent in the northern summer; the situation is reversed in the southern hemisphere, where the summers are 3 per cent warmer and the winters 3 per cent cooler than they would be if the Earth's orbit round the Sun was strictly circular.

This situation is constant for several centuries, but over several thousand years there is a marked change arising from the gravitational influence of the Sun and Moon as they affect the axis of spin of the Earth. This effect, discovered by the ancient astronomers, and first calculated by Isaac Newton, is remarkably similar in its mathematical properties to the precession of an ordinary spinning top. Just as a top spins about its axis, which moves in a cone around the vertical direction, the Earth's axis of spin moves in a cone around a direction that is perpendicular to the plane of its orbit, the ecliptic. This motion is shown schematically in Figure 17A. Because of the precession of the Earth's axis

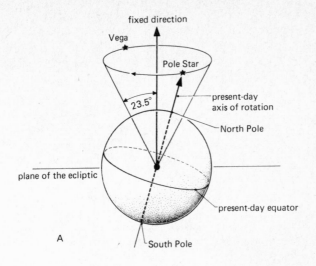

A

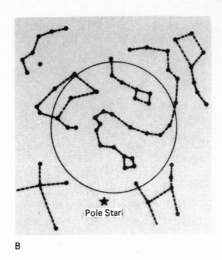

B

Figure 17: *Precession of the Earth's axis of rotation. The Earth's axis of rotation precesses like a spinning top (A), so that the direction in the sky towards which the axis points moves around a circle (B). At present, it points close to the Pole Star, but will move around the circle through the constellations shown to return to its present position in about 26,000 years.*

63

of spin, the direction towards which the axis points among the stars changes. The axis goes slowly round and round the circle shown in Figure 17B, completing each circuit in about 26,000 years. Figure 17B shows the present well-known pointing of the axis towards the Pole Star. In another 13,000 years (about 15,000 AD), the axis will be halfway round the circle, far away from the present Pole Star, pointing close to the bright star Vega.

The precession of the Earth's axis of spin changes the times of year when the Earth is nearest and farthest from the Sun. In the middle 1860s, James Croll calculated the effect of this precession on the Earth's climate (Table 2).

Table 2: **Effects of precession of the Earth's axis of rotation on summers and winters**

time (years before present)	northern hemisphere	southern hemisphere
present	summers 3% cooler winters 3% warmer	winters 3% cooler summers 3% warmer
13,000	summers 3% warmer winters 3% cooler	winters 3% warmer summers 3% cooler
26,000	summers 3% cooler winters 3% warmer	winters 3% cooler summers 3% warmer
39,000	summers 3% warmer winters 3% cooler	winters 3% warmer summers 3% cooler

and so on, alternating at intervals of 13,000 years

As shown in Table 2, the precession of the Earth's axis of rotation, taken by itself, causes a switch every 13,000 years in the summer–winter asymmetry in the supply of solar heat to the northern and southern hemispheres. This is on the assumption, however, that no change takes place in the orbit of the Earth around the Sun. Small gravitational effects caused by other planets actually produce slow changes in the orientation of the Earth's orbit, which alters the switches of Table 2 from 13,000 years to about 11,500 years. James Croll calculated these results without the aid of any mechanical or electronic calculator, and they accord remarkably closely with modern theories. Croll's achievement was all the more remarkable in that he had little formal education, and that his job was the modest one of janitor to the Andersonian College and Museum of Glasgow. His work came as a considerable surprise to the academic world, and it says much for the social broadmindedness of Victorian Britain that, from 1867 onwards, James Croll was accepted as a scholar of distinction by the scientific world.

If one attempts to argue that ice ages are caused by the changes shown in Table 2, there are difficulties – difficulties that were not apparent in Croll's lifetime, but which are only too obvious today. Since there is no present ice age in the northern hemisphere, one cannot associate ice ages with cooler summers. The association would need to be with cooler winters – that is, there should have been ice ages in the north at 11,500 BP, at 34,500 BP, at 57,500 BP, and so on. Interspersed between these ice ages, there should have been a comparative absence of northern ice: an absence at present, at 23,000 BP, at 46,000 BP, and so on. This is evidently in conflict with the facts, since 23,000 BP was close to the height of the last ice age. Moreover, the asymmetry between the north and the south requires an alternation of ice ages in the two hemispheres, which is also wrong. Modern evidence shows ice ages in the north and the south to have been contemporaneous. This first attempt at an astronomical theory of the ice ages was therefore unsuccessful.

One way to attempt to breath new life into an unsuccessful theory is to make it more complicated. While such a procedure is rarely successful ultimately, it may attract some notice, especially if the details are made sufficiently awkward to understand. This was the procedure followed with respect to Croll's theory by Milutin Milankovitch. To understand the complication introduced by Milankovitch, let us begin with the simple observation that the seasons of the year are caused by the changing angle between the Earth's axis of spin and the direction of the Sun.

As can be seen in Figure 16, when this angle is least (for our hemisphere) we have midsummer; when the angle is greatest we have midwinter. If the Earth's polar axis had happened to be exactly perpendicular to the plane of the Earth's orbit around the Sun (the ecliptic), then of course the angle which it made with the direction of the Sun would always have been 90° and there would have been no seasons. Every day would have been like every

65

other day. If, on the other hand, the axis of spin were turned so as to make a zero angle with the ecliptic, the seasons would be of a ferocious intensity that we can scarcely imagine. Climate on the Earth is tuned therefore to the particular angle of tilt of the Earth's spin axis that we happen to have. This angle, currently about 23.5°, changes with each millenium to a small extent.

Figure 18 shows the variations of tilt angle that are calculated to have occurred over the past 250,000 years. According to the calculations, the tilt angle has varied between a low value of about 22° and a high value of about 24°. Milankovitch claimed that the resulting fluctuations of climate were big enough, in spite of the small range of the angle, to be a major factor in the cause of ice ages.

Some years ago I had reason to work out mathematically which place on Earth receives the most solar radiation at midsummer. I found to my astonishment that it was the pole. It is obviously not the pole that receives most radiation at midday but, because the Sun never sets at midsummer for latitudes inside the Arctic circle, sunlight is received there for the full twenty-four hours.

In making this calculation, I assumed there was no cloud, water vapour or dirt in the atmosphere. I also omitted the scattering of sunlight by air molecules – that is, the scattering which produces the blue colour of the sky. If scattering were included, it would be stronger at the pole than at the equator, because the lower elevation angle of the Sun at the pole causes the light to pass through a greater air mass there than at the equator. The

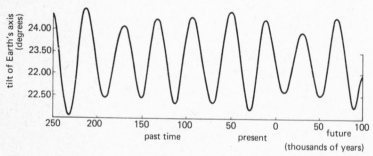

Figure 18: *Variations in the tilt of the Earth's axis.*

effect at the pole is to reflect more of the sunlight back into space, reducing the radiation received by about 10 per cent relative to the equator. The inclusion of scattering would therefore have modified my calculation, but not sufficiently to change the remarkable result that for a cloud-free, dust-free atmosphere it is the pole which receives most solar radiation on midsummer day.

For the purpose of comparing the radiation received at slightly different angles of tilt of the Earth's axis of spin, we can ignore the complex refinement of scattering by air molecules, because the scattering will not change significantly over the small range from one angle of tilt to a closely similar angle. Standardizing to the radiation received at the pole when the tilt angle is 23°, one calculates the curves of Figure 19 for the radiation received on midsummer day at latitudes from 50° to the pole, and for the

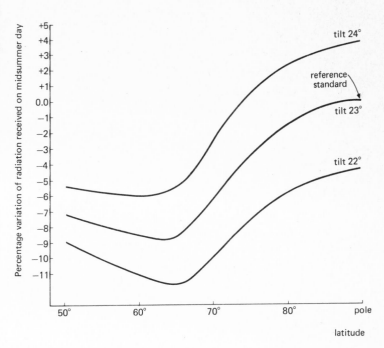

Figure 19: *Percentage variation in solar radiation at midsummer day, received according to tilt angle and latitude. The reference standard is the pole when the tilt angle is 23°.*

three tilt angles of 22°, 23° and 24°. This is the maximum range of the Milankovitch effect. At latitude 65°, typical for ice-sheet formation, the fluctuation of the 24° and 22° curves from the curve for 23° is about ±3 per cent, very much the same as the Croll effect of Table 2.

The Croll effect behaves in opposite ways in the northern and southern hemispheres, whereas the Milankovitch tilt effect is the same in both hemispheres. The two effects differ also in their periodicities: the oscillation time for the Milankovitch effect is about 41,000 years (as shown in Figure 18), almost twice as long as the oscillation time of 23,000 years for the Croll effect. Trying to reconcile the two effects may therefore be expected to lead to a complex situation. Table 3 summarizes the results for three epochs: the present, 13,000 BP, and 23,000 BP. The percentages given in the table represent changes from the long-term average of the radiation received over the three months from early May to the end of July.

Table 3: **Summary radiation received at latitude 65°**

epoch	northern hemisphere	southern hemisphere
present	1% cooler than average	3% warmer than average
13,000 BP	4% warmer than average	average
23,000 BP	4% cooler than average	average

A recent article by J. D. Hays, J. Imbrie and N. J. Shackleton claims that the astronomical theory of the cause of ice ages has been proved to be correct.[1] The methods used by these authors have been very clearly discussed in the book *Ice Ages: Solving the Mystery* by J. Imbrie and K. P. Imbrie.[2]

The use of deep-sea cores for determining the time variations of the amount of ice on the land is explained in Technical Note 3. It is possible to analyse these time variations (which in themselves look very complex) by a delicate statistical technique. The aim of the statistics is to discover which oscillation periods will

[1] *Science*, Volume 194, 1976, p. 1121.
[2] Macmillan, 1979.

generate, when added together, the complex form of the experi-
mentally determined variations. For the Croll effect, there are
two main oscillation periods: one of about 23,000 years and the
other of about 100,000 years. (The latter period arises from
changes in the gravitational effects of the planets on the Earth's
orbit around the Sun.) For the Milankovitch effect, there is an
oscillation period of about 41,000 years, determined by the
variations of tilt of the Earth's axis of spin (see Figure 18). What
Hays, Imbrie and Shackleton find is that these three periodicities
– 23,000 years, 41,000 years and 100,000 years – all appear in a
significant way when statistical analysis is applied to the time
variations revealed by their deep-sea cores. From this, the three
scientists conclude that the ice ages must have been caused by the
Croll and Milankovitch effects. I question the logic of this
conclusion.

If an ice age was brought about in some quite different way
from that claimed by Milankovitch, the ice sheets on the land
would be subject to variations caused by the Croll and
Milankovitch effects. The variations might be small, but they would
occur. Other things being equal, the ice sheet would extend itself
a little if the summers were cool, and it would retreat a little if the
summers were warm. An ice sheet existing in the first place as the
result of something quite different would therefore become
'modulated' by the Croll and Milankovitch effects. With a
sufficiently sensitive statistical technique, it would be possible to
pick up the modulations. From this simple consideration, one can
see that the evidence of the modulations does not prove anything
about the primary cause of the ice ages.

I am also doubtful about the method itself. The time sequences
in these deep-sea cores are not unequivocal, as they would be if
the various depths in the cores could be dated by the use of the
radioactive-carbon method, (unfortunately the radioactive-
carbon method for determining time cannot be extended far
enough back into the past). The ingenious magnetic method
used by Hays, Imbrie and Shackleton explained in Technical
Note 2 is necessarily coarser than a direct radioactive method
would have been, and adjustments involving human intervention
(which Hays, Imbrie and Shackleton refer to as a 'tuning' and
'patching' of the data) had to be made.

Apart from these reservations, there are further important reasons why I do not believe the Croll and Milankovitch effects could have caused the glaciations of the past million years or so. The calculations of James Croll (confirmed in modern times by the Belgian astronomer André Berger) show that the variations from the average arising from the Croll effect are generally in the region of 7 per cent.[3] The Milankovitch effect, however, has never been much greater than 3 per cent, even on midsummer day, and so over most of the past 500,000 years the Croll effect has been dominant. Thus, over a broader time interval than that of Table 3, variations in the tilt of the Earth's axis of spin have been only an addition to the Croll effect. On a broader time base, therefore, the Milankovitch theory is essentially the same as the Croll theory. If it were correct, ice ages should alternate in the northern and southern hemispheres, but the evidence from oceanic cores shows that ice ages occur contemporaneously in both hemispheres.

Another flaw in the Croll and Milankovitch theories lies in their long periods of oscillation: 23,000 years for the Croll effect and 41,000 years for the Milankovitch effect. Both are slowly varying and ponderous; they could never explain the sudden warming and subsequent cooling which occurred with the extinction of Neanderthal man (see Chapter 2), or the sudden flash of warm climate followed by a relapse to cold conditions at about 13,000 BP which foreshadowed the end of the last ice age. These sudden, sharp pulses of warmth occurring in a few decades evidently demand a very different kind of explanation – one that allows for drastic changes of climate which occur in only a few years.

The Milankovitch theory contains other inconsistencies. For example, slight changes in the tilt of the Earth's axis of spin give only negligible solar variations at equatorial latitudes; yet the last ice age produced the great glaciers on Mauna Loa and Mauna Kea in Hawaii, and Mount Elgon in Uganda. Obviously, something drastic happened in the tropics for which the Milankovitch theory cannot account.

[3] The variation at present is 3 per cent, as shown in Table 2, but taken over a longer time scale of 500,000 years the variation is about 7 per cent.

70

It is also unable to explain why a 4 per cent warming at 13,000 BP in the northern hemisphere was able to melt the northern glaciers in the exceedingly short time of a century or two, and yet the current 3 per cent warming in the southern hemisphere cannot manage to melt the Antarctic ice at a rate of more than 1 millimetre a year. It could be suggested that the Antarctic ice sheet is so vast that it is proof against the present 3 per cent warming of the southern hemisphere but, if that is so, how did a great quantity of Antarctic ice disappear at the end of the last ice age in spite of there being no warming at all in the south at that time?

Comparing Milankovitch's calculations with evidence shown in Figure 8, it would appear that, if his theory were correct, the glaciers which reasserted themselves at 11,000 BP must have been caused by particularly cold winters, as Milankovitch himself believed. But climatologists and glaciologists have generally agreed that winters are largely irrelevant. Temperatures are low enough in winter for snowfalls to persist over the relevant areas of the world. As shown in Figure 20, temporary snow lies in all the areas that were glaciated in the last ice age. Evidently there is no problem about the winters – snow always falls and remains through the winter. The issue is whether or not the snow melts during spring and summer.

Snow is highly reflective of sunlight. Indeed, pristine snow reflects about 90 per cent of incident sunlight, which means that the Sun is powerless to melt clean snow. (Snow does not melt on a spring day in the British Isles because of the Sun; it melts because of warm air from the sea.) After snow has packed down into ice, a different problem arises. Clean ice absorbs about two-thirds of the sunlight incident upon it, but ice is sufficiently transparent for the light to penetrate quite a long way (10 metres or more) before absorption takes place.

It is remarkable what profound results follow from this simple property of transparency to sunlight. If, instead of penetrating deeply, the light were absorbed in a shallow surface layer of ice, the summer sun would quickly raise the temperature of the thin surface layer to melting point, and almost immediately water would run off. But, when the sunlight penetrates a thick layer of ice before it can be absorbed, the sunlight cannot raise the

71

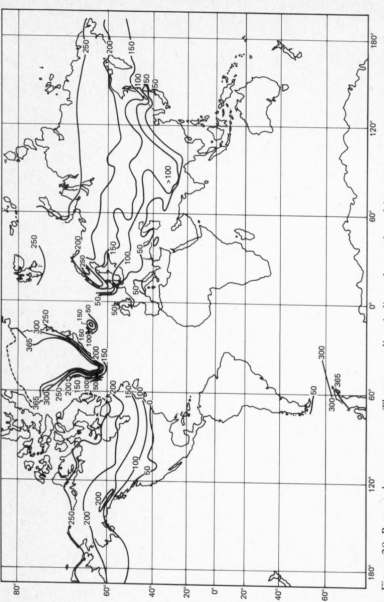

Figure 20: *Present-day snow cover. The contour lines indicate the number of days in an average year that snow lies on the ground. Comparison with Figure 1 (page 18) shows that snow lies nowadays in essentially all the areas that were covered by ice sheets 20,000 years ago. (After H. H. Lamb, Climate: Present, Past and Future, vol. 1, Methuen, 1972.)*

temperature of the ice to melting point quickly enough. When the ice is very cold, the whole summer passes before any melting occurs at all. This is what happens today in the Antarctic, just as it must have happened in northern Europe during an ice age. (It is interesting to reflect that, if by some magic ice were suddenly made opaque to light, the glaciers that exist today would melt away in a few years, raising sea level by 60 metres or more, and so inundating at least half of the world population. So much can depend on so simple a physical property.)

The average annual precipitation over Antarctica is surprisingly low – only 15 centimetres water equivalent, providing about 1.5 metres of fluffy snowfall. It seems astonishing that such a low precipitation rate could be enough to feed the huge Antarctic glaciers such as the enormous Byrd Glacier referred to in Chapter 3. The reason it can do so lies in the exceedingly low rate of evaporation and melting of Antarctic snow and ice. In attempting to explain this remarkable fact, it would be natural to argue that the low melt rate is partly due to low temperature and partly to lack of sunshine. The first of these points is true and relevant but, as Figure 19 shows, the notion that there is little solar radiation at high latitudes even at midsummer is illusory; there is more radiation in polar regions than at the equator. Even though in the southern summer the Antarctic receives more radiation than anywhere else on the Earth, the melting of the Antarctic ice is confined to a mere millimetre per year. Yet the Milankovitch theory suggests that a change of a few per cent in summer radiation can cause the Earth to plunge from the comparatively warm conditions we enjoy today into the depths of an ice age.

The size of the modern world's greatest ice sheet and the glaciers that flow from it in Antarctica is not regulated by summer melting. It is regulated by the ice dropping off the ends of the glaciers when they reach the sea, producing icebergs. It is estimated that this 'calving' process is at least a hundred times more effective than melting and evaporating in limiting the Antarctic ice. Since about 90 per cent of the modern world's ice is in Antarctica, what happens there is far more important than what happens elsewhere. Most of the remaining 10 per cent of the world's ice is in Greenland, not in the Alpine glaciers studied so avidly by the early pioneers. It is now thought that in Greenland about half the ice that is 'lost' goes in the calving of icebergs and half in summer melting.

There is one particular situation, sometimes important for mountain glaciers in middle latitudes, in which sunshine can cause an appreciable melting of ice. When the temperature of the ice is not much below its melting point to begin with, even though the sunlight still penetrates some 10 metres into the ice, only a little energy is required to lift the temperature of the whole surface layer to its melting point. Under these conditions, a strong summer sun shining over several weeks can provide sufficient absorbed energy to convert such a layer of ice into mush, often with spectacular effects. Besides producing swift-flowing glacier streams, water from the mush may accumulate into temporarily dammed-up lakes within a glacier. The water in such lakes is under great pressure from the weight of the overlying ice and, if the pressure is suddenly released, the water spouts upwards from inside the glacier.

A more powerful effect of a similar kind occurs when the ice at the bottom of a glacier is melted by volcanic heat. Very large water bodies can then accumulate, as in the case of Grimsvotn on the Vatnajökull in Iceland. Every ten years or so, 'glacier bursts', known as *jökull laups*, cause the large trapped lakes to break out with effects as catastrophic as those of the collapse of a man-made dam. In the Grimsvotn the bodies of water so released are large enough to flood several hundreds of square kilometres of land in the valley below the glacier. In these bursts, vast blocks of ice are carried far beyond the normal range of the glacier, and this might be one way in which erratic blocks (see Chapter 3) could have been distributed at the end of the last ice age. Boulders encased within ice blocks could have been exploded by water pressure out of the glaciers of the Lake District, Scottish Highlands or Norwegian mountains, and the ice blocks might then have floated many miles on temporarily dammed-up lakes. This probably occurred, but it cannot explain all the erratics that have been found. Boulders could not have been floated uphill, and it seems impossible that erratics from very small eminences like Ailsa Craig, or the Shap intrusion, could have been broadcast in this way.

Melting by the Sun of initially not very cold ice occurs in the Alps and in the north-west Pacific, but not under polar conditions, which are the best approximation we have today to the conditions of an ice age. Yet, even in the Alps, warm air is a far more devastating cause of melting than sunshine. I cannot do better by way of illustrating this point than to quote Frank S. Smythe, outstanding among the pre-war generation of mountaineers. He had been crossing the Clariden Pass to the Mederanerthal one spring:

So far it [the weather] had been quiescent with a suspicion of frost in it, but now I felt light puffs of wind against my cheek and puffs of astonishing warmth. It was as though I was standing in the cold on the threshold of a house and someone opened the door letting out the warmth of the house into my face. It was the Föhn wind. . . .

[Before long] I heard a sharp hissing sound like an angry serpent. Two hundred feet above me an avalanche was starting. A little coverlet of snow resting on a rock slid off on to the slope beneath; it set a small ball of snow rolling; this gathered mass and impetus; it attained the dimensions of a cartwheel, then burst to form into fragments; each fragment augmented itself, then burst to form other fragments. It was only a matter of a second or so for this to happen. . . . I remember saying out loud: 'Now you're for it.'

In fact, he reached the safety of a hut from where he could observe the effect of the warm Föhn wind:

Avalanches were falling every few seconds and within an hour the sound of them was so continuous that not a moment's silence was detectable. There must be many who wondered how the enormous accumulations of winter snow are cleared away; they would not do so if they had seen that afternoon's work. I had only to close my eyes to imagine myself in the midst of some Alpine battle area, for never have I listened to such a volleying and thundering. From the cliffs of the Düsselstock alone there was a continuous fusillade of minor snow avalanches and ice fragments, and farther down the valley larger masses were falling over the line of cliffs above the steeply sloping terrace along which I must eventually go to reach the Maderanerthal, sweeping the terrace and finally pouring over the lower cliffs on to the floor of the valley. Frozen waterfalls augmented by winter frosts into columns of ice hundreds of feet high were torn loose and every few moments the slopes above the upper cliffs slid bodily to destruction over the cliffs in cataracts that must have weighed many thousands of tons. Even greater were the monstrous ground-avalanches that fell into the Maderanerthal from the Oberalpstock. . . .[4]

The heat of the Föhn wind comes from the condensation of water vapour from the oceans into liquid droplets. It takes a lot of energy to convert a kettleful of boiling water into steam; simply raising the temperature of water to boiling point is not remotely sufficient to change the liquid into vapour – you have to go on maintaining the heat under a kettle for a long time to do that. Indeed, the energy required to convert a quantity of boiling water into steam is about five times greater than the energy needed to raise the temperature of the same quantity of ice-cold water to boiling point. The great amount of energy required to evaporate liquid into vapour is not irrevocably lost. It can be recovered by condensing the water vapour back into liquid, and it is this return of the energy, originally supplied in the evaporation of vapour from the oceans, that produces the warm air of the Föhn wind.

The reason for the importance of warm air in melting ice and snow lies in basic physics. Sunlight is an inefficient melting agent because it is strongly reflected by snow and it penetrates deeply into ice. Heat rays, on the other hand, which are radiation of very much longer wavelength than visible light, behave in the opposite way. Heat rays are not strongly reflected by either snow or ice, and heat radiation penetrates only a little way before it is absorbed. Consequently, the energy of heat radiation goes into a thin surface layer, which therefore rises quickly in temperature. For snow, the melting when warm air comes occurs almost before your eyes. There is probably no place in the world where it is easier to see the power of the oceans to melt snow and ice than

[4] Frank S. Smythe, *An Alpine Journey*, Hodder and Stoughton, 1934.

the British Isles. Sunny but cool days have little effect, unless the snow becomes very dirty. But with the first breath of really warm air from the Atlantic the snow is gone. For ice, which is denser than snow, the melting takes longer, but even a considerable depth of ice will melt off in a day or two in warm air.

The oceans serve the Earth as a night-storage heater serves a house – a huge night-storage heater indeed, for the amount of energy that could be taken out before the sea would freeze is as much as the Sun supplies to the whole Earth in a decade. This simple fact makes nonsense of the astronomical theories of the cause of ice ages. Neither the Croll effect nor the Milankovitch effect changes the total amount of solar energy received by the Earth. The Croll effect unbalances the two hemispheres a little – one hemisphere gains a few per cent extra energy at the expense of the other hemisphere, with the gains and losses alternating every 11,500 years or so, and the Milankovitch effect makes about a 1 per cent change in the distribution of solar energy between polar and equatorial regions.[5] If climate were strictly localized, such changes might be of some significance, but climate is world-wide. Roughly half of the energy which heats the regions of the Earth poleward of latitude 50° comes from water vapour that evaporated from the surface of tropical areas – that is, 50 per cent, rather than the negligible 1 per cent of the Milankovitch effect.

If I were to assert that a glacial condition could be induced in a room liberally supplied during winter with charged night-storage heaters simply by taking an ice cube into the room, the proposition would be no more unlikely than the Milankovitch theory.

[5] The maximum change on midsummer's day at latitude 65° is 3 per cent. The maximum change at latitude 65° for the three months from May to July is 2 per cent. Then, because of the small area of the polar cap, the average effect for the whole hemisphere falls to less than 1 per cent.

5 The Cooling of the Primordial Earth

If the often-discussed astronomical theories of the cause of ice ages are unacceptable, the problem remains of what the real cause was – not only of our own glacial epoch (say, from 5 million BP), but of the great Permo-Carboniferous (340–240 million BP), the late Ordovician (460–410 million BP), the Upper Proterozoic (1200–600 million BP) and the Huronian (2300 million BP). To find the solution, we must look back far beyond the earliest glacial epoch to the dark and sinister world of the Archean (from the origin of the Earth, about 4500 million years ago, to about 2500 million BP), when the Earth was covered by a torrid ocean and when vast cloud banks obscured the Sun.

Let us start with a puzzle. Albert Einstein's generally accepted theory of gravitation holds that the energy output of the Sun has slowly increased, by about 30 per cent, since the Lower Archean, and yet no glacial epochs are known until 2300 million BP. If a drop of 30 per cent in the energy output of the Sun were to occur today, the average temperature of the Earth would decline by at least 30°C (about half of this reduction being a direct effect of the lower intensity of sunlight, with the rest coming from a variation of the cloud and greenhouse effects discussed later in this chapter). Such a cooling would not only cause ice sheets to form on the land, but would make all the oceans freeze solid. Yet most Lower Archean rocks, whether in Canada, Greenland, Australia or South Africa, are either sedimentary (i.e. laid down in water), or igneous rocks from volcanoes, which have a kind of blistering that is characteristic of hot lava being suddenly cooled by injections into water. This suggests that, although there was some *terra firma* in the Archean, there was not a great deal of it, and that the Archean ocean probably covered almost the whole world. Moreover the ocean was liquid.

If the early Earth had been cold, there would have been very little water vapour in the atmosphere and consequently little cloud. This is not, however, the opinion held by many geologists. Dr G. Young has written that:

> The Pre-Cambrian world would have been largely unrecognizable to us. . . . For much of the time, at least during the early part of the Pre-Cambrian (Lower Archean), the Sun was hidden by unbroken banks of cloud. It rained – possibly for millions of years – without a break.[1]

Some years ago, when I was trying to reconcile the conflict between the evidence suggesting that the Earth was much hotter 3000 million years ago than it is today and Einstein's theory that the Sun was then 30 per cent less luminous than it is now, I heard of a group of geochemists who had determined the temperature of the ocean about 3000 million BP, using a modification of the O–16 and O–18 method discussed in Technical Note 3. The work had not been published and so I made a special journey to see it. There was no question of its technical competence; the one uncertainty in it was that the ocean had been taken to have the same oxygen and hydrogen isotope ratios then as it has now.[2] Since the Archean ocean covered virtually the whole world, most of the water in the present ocean must have been there then, in which case the assumption of similar isotope ratios could not, I think, have been seriously wrong. The result obtained by the geochemists for the temperature of the ocean of 3000 million BP was 50°C – that is, about 30°C hotter than the modern ocean. At such a remarkably high temperature there would have been enormous evaporation and precipitation, just as Dr Young suggests.

The earliest forms of life on our planet (Figure 21) were those which could withstand very high temperatures: bacteria that can exist and thrive up to the boiling point of water, and blue-green

---

[1] G. Young, *Winters of the World*, David and Charles (1979).
[2] The oxygen isotopes are the O–16 and the O–18 atoms of Technical Note 3. The hydrogen isotopes are usually referred to as ordinary (light) hydrogen and deuterium (heavy hydrogen).

algae that can exist up to about 75°C. Many people must have wondered why it was so long before multi-celled creatures appeared on Earth and why there was little in the way of evolution of life over the vast interval of 3000 million years from the Lower Archean to the Upper Proterozoic. The reason is clearly that the early Earth was so hot that nothing but high-temperature life-forms could exist. It was only with the marked cooling of the Upper Proterozoic that more complex structural forms could begin to develop.

The only factor that can explain the apparent contradictions between Einstein's theory and the evidence of the Earth's high temperature 3000 million years ago is that the early terrestrial atmosphere was crucially different from our own. I believe that the ancient atmosphere contained a very large amount of carbon dioxide, perhaps 30,000 times more than that in our present atmosphere.

The first evidence for this comes from carbonate rocks:

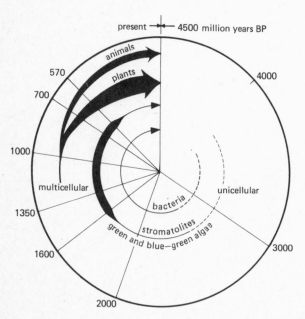

Figure 21: *Life forms in the geological record.*

limestones, dolomites and deposits of coral. The carbonate rocks were not primordial to the Earth; they have all been formed by calcium and magnesium oxides acquiring carbon dioxide, and the amount of carbon dioxide needed to produce all of them is many thousands of times greater than the amount in the atmosphere now. I believe that the only reasonable conclusion is that the carbon dioxide required by the carbonate rocks was present in the Archean atmosphere.

According to my possibly oversimplified view, both the carbon dioxide and the water must have been present at the terrestrial surface ever since the oldest known surface rocks were laid down – that is, 3800 million years ago. The popular view among geologists, however, is that both the water and the carbon dioxide were somehow buried inside the primordial Earth, and that both have since been outgassed by volcanic action. If this were so, the outgassing of the water must have occurred quickly because the ocean had covered the terrestrial surface by 3800 million BP and, if outgassing were quick for the water, it must also have been quick for the carbon dioxide. The outgassed carbon dioxide must have accumulated in the atmosphere throughout the Archean, since no appreciable quantities of carbonate rocks were formed before the *end* of that epoch. Clearly, the amount of carbon dioxide in the early atmosphere must have been very large.

The carbon dioxide would have created a radiation trap near the Earth's surface – a trap not for light but for the long-wavelength heat rays emitted by the Earth and by the atmosphere itself.[3] Such radiation traps occur in man-made greenhouses, from which they take the name 'greenhouse effect'. Carbon-dioxide traps are much in the news these days because of widely disseminated stories about the consequences of burning the world's coal reserves. In fact, these stories exaggerate the effect of the addition of carbon dioxide to the atmosphere, because they overlook a crucial physical aspect of the problem, namely that the Earth is a water plant with a clever way round heat traps, which a dry planet such as Venus does not have.

One can feel reasonably comfortable even on an exceedingly

[3] For a discussion of the concept of 'wavelength', see Technical Note 4.

hot day – say, when the temperature is 120°F (50°C), well above the body temperature of 97°F (37°C) – as long as the air is very dry. There is no impression of sweating on a dry day, because perspiration at the skin can evaporate immediately. When the air is very hot and humid, perspiration does not evaporate easily and the body becomes uncomfortably 'lathered' in sweat. Indeed, at 100 per cent humidity, one could not live at all in air temperature much above 97°F (37°C), because there would be no way to lose the excess heat. A few minutes in such conditions would be enough to cause fatal 'heat stroke' – that is, heart failure.

I have often thought it odd that there is no place on Earth where this lethal condition occurs, although the climate in summer on the east coast of the United States sometimes comes quite close to it. The reason for this 'near-miss' cannot lie with the Earth, of course; it must be a consequence of biological evolution. Our heat-loss mechanism evolved to cope with the worst summer conditions to be found on the Earth. This observation implies that the worst summer condition cannot have changed much over the past few million years, since an environmental factor has to remain steady over a time scale of this order if biological evolution is to be able to respond to it.

The Earth copes with heat-radiation traps in much the same way as the human body. Evaporation of water vapour at the ocean surface cools the ocean. The water vapour rises with updrafts of air, producing clouds. When the clouds are big enough – for example, the huge anvil-shaped cumulus clouds seen today – a fraction of the water vapour is carried above the radiation trap. That water vapour condenses into droplets, or into ice crystals, and the energy that was latent in it is released into the atmosphere. Because this happens above the radiation trap, the released energy is emitted into space, and cannot be reabsorbed.

We are in a position to make a very interesting calculation. If the heat trap in the Archean atmosphere was entirely efficient, no *radiant* heat at all would have leaked through it from the ocean surface into space. Dr Young's 'unbroken cloud banks' would have produced a dark, lowering effect, allowing only a fraction of sunlight – perhaps 20 per cent – to get through to the surface. I will suppose this fraction of penetrating sunlight to have been equal to the fraction of water vapour that was carried upwards in rising columns of air to heights above the trap. This supposition of equality has the advantage that we do not then need to bother with an explicit specification of the fractions, because they simply cancel each other out in the calculation. It is then easy to show that the oceans will heat up as a result of heat trapping until the average rate of evaporation (and therefore the average rainfall over the Earth) rises to about 4 metres of water per year, which is some five to six times the average rainfall of the whole Earth today. In other words, during the Archean epoch, the oceans would have heated up until the evaporation rate was about six times faster than it is at present.

It is now possible to confirm the temperature of the Archean ocean. Assuming that the ocean today evaporates at an average temperature of 20°C, the saturation water-vapour content of the air at the ocean surface is equivalent to a barometric pressure of 1.75 centimetres of mercury. For a saturation water-vapour pressure six times greater than this (i.e. for a barometric pressure of $6 \times 1.75 = 10.5$ centimetres of mercury, the evaporation rate increases sixfold. Water-vapour tables in any physics or chemistry reference book will verify that the water temperature required to give a vapour pressure of 10.5 centimetres of mercury is 53°C. (Actually these figures are for pure water. A modest correction upwards would be needed for salt water, but the correction would be no more than a degree or two.) The coincidence of this calculation with the geochemical determination of the temperature of the Archean ocean is impressive.

It must therefore have been a powerful carbon-dioxide heat-radiation trap in the atmosphere that generated the high temperature of the Archean, thereby preventing life from evolving beyond the stages of algae and bacteria.

To a space-borne observer during the Archean, the Earth would have seemed locked into a fixed state. Hundreds of millions of years rolled away with no apparent change. Yet all through that vast timespan the volcanoes on the sea floor were at work. As rocks were heated and reheated, the dispositions and arrangements of minerals within them were changing. In time, a form of rock with an exceptionally high proportion of silica accumulated in greater and greater quantities. It was a kind of geochemical rubbish but, having about 20 per cent less density than the rocks typical of the outer shell of the Earth, it tended to ride like a cork in the sea. The rising areas of low-density material were to become the cores of the continents – the Archean shields that can be seen today in Greenland, Canada, South Africa and Australia and the north-west of Scotland. As the cores rose out of the sea, underwater shelves were formed around them, providing a habitat for blue-green algae. With the resulting proliferation of blue-greens, the Earth emerged from the dark and sinister epoch in which it had for so long been locked.

Blue-green algae feed on carbon dioxide, using sunlight and small concentrations of trace materials dissolved in water to turn carbon dioxide into sugars. Then, using the energy stored in the sugars, the algae build all the complex biochemical substances that they require to replicate themselves. Replication is limited only by the availability of carbon dioxide in sunlit areas of shallow water containing sufficient supplies of dissolved trace materials. In perfect conditions, the replication rate becomes enormous – a single cell could in a week or two generate a colony with a mass equal to the whole Earth, and in a month or so a mass equal to the whole universe – and such 'primitive' cells are potentially immortal, dying only in adverse conditions. Towards the end of the Archean, conditions did become right. Indeed, I would look on the proliferation of blue-green algae as the event which really marked the end of the Archean, rather than anything that profoundly affected the crustal rocks of the Earth.

Until then, the small amount of oxygen in the atmosphere had come from a slow dissociation of water vapour caused by solar ultraviolet light. But that quantity was insignificant compared to the flood of oxygen provided by the blue-green algae. For each two carbon-dioxide molecules they used, they released into the atmosphere one molecule of oxygen.

The following picture emerges. A dramatic biological explosion in the number of blue-green algae would have removed carbon dioxide almost completely from the atmosphere, replacing the carbon-dioxide molecules on a one-for-two basis with oxygen molecules. Because oxygen does not generate a significant heat-radiation trap, the temperature of the Earth would have fallen sharply. Indeed, because the Sun was still 15 per cent less luminous than it is now, there would have been an

immediate glacial epoch. This coincides with the Huronian glaciation shown in Figure 5 (page 25).

While I think this simple picture is broadly correct, it would be
impossible for a single biological explosion to remove all the
atmospheric carbon dioxide in one fell swoop. The blue-green
algae could multiply only to the extent that their habitat permitted; they would still have been limited by the supply of trace
elements dissolved in the water and the available area of sunlight. Failures in the nutrient supply would have caused some
algal cells to die. The dead cells would then have fallen slowly to
the shallow ocean floor, building up into a store of dead biomaterial analogous to the inundated fern forests of the Permo-
Carboniferous which gave rise to the coal measures that we
exploit today.

There is another process which would have slowed down the
rate of abstraction of carbon dioxide from the atmosphere. If the
carbon taken from the present-day atmosphere were irreclaim-
ably stored in living and dead biomaterial, our supply of carbon
dioxide would be lost very quickly. Indeed, the present-day
reserve of atmospheric carbon dioxide would disappear in only a
few years. This does not happen, because dead biomaterial
returns its carbon dioxide to the atmosphere, whence it is re-used
by living organisms. Living plants abstract carbon dioxide and
add oxygen to the atmosphere, while dead plants abstract oxygen
and return carbon dioxide. Year after year sunlight stimulates
the photosynthetic plants to continue this cycle.

It is possible to calculate how much carbon dioxide was abstracted from the atmosphere by the blue-green algae from an
examination of the Earth's crustal rocks. The rocks of the Earth
are primarily condensates from a gas at high temperature – about
1000°C – although some chemical interchanges have occurred at
temperatures between 1000°C and our present low temperature
– interchanges at perhaps 500°C.

Iron provides an important example. At high temperatures,
iron condenses in its metallic form; while, at intermediate temperatures, in the presence of water, it takes non-metallic forms. It
is most stable then as ferrous oxide with some iron sulphide. It is
not until rocks are processed at significantly lower temperatures
– less than about 150°C – that the preferred form becomes the

more highly oxidized ferric iron. The iron in the Earth's crustal rocks occurs overwhelmingly as ferrous oxide, not as ferric oxide. Yet, after oxygen had been generated in the atmosphere by the blue-green algae, there was a slow but steady conversion of ferrous oxide into ferric oxide. Ferric oxide is bright red, and it was the oxidation of ferrous to ferric that caused most of the red sands and soils of the Earth.

If we suppose all the ferric oxide now found in the crustal rocks of the Earth to have been formed through primordial ferrous oxide taking up atmospheric oxygen, it is simple to estimate how much oxygen must have been supplied by the blue-green algae. The crustal rocks for which the chemical composition is reasonably well known have an average thickness of 10 kilometres, so that, for each square metre of the Earth's surface, there are about 30,000 tons of crustal rock, of which about 0.5 per cent is known to be ferric oxide. In other words, there are 150 tons of ferric oxide per square metre of the Earth's surface. Nine-tenths of this weight would come from the original ferrous oxide itself and only one-tenth, 15 tons per square metre, from the absorption of atmospheric oxygen. The current amount of atmospheric oxygen is about 2 tons per square metre, so that the amount of oxygen produced by the blue-green algae had to be some eight times the present amount of oxygen, rather more than all the nitrogen of our atmosphere. The corresponding weight of the original carbon dioxide would need to have been about three times greater still; four to five 'atmospheres' of carbon dioxide would have been required to generate all the oxygen that changed ferrous to ferric iron.

This takes no account of the carbonate rocks mentioned above. The amount of such rocks in the Earth's crust has not yet been finally calculated but, taking an average of the various estimates, it seems that the carbonate rocks required about half as much carbon dioxide as the estimate of the previous paragraph. Taken together, the ferric oxide and the carbonate rocks required five to ten 'atmospheres' of carbon dioxide – that is, five to ten times the amount of the present-day atmosphere. This estimate is unconventionally large, but it is still less than 10 per cent of the amount of carbon dioxide in the atmosphere of the planet Venus.

86

The blue-green algae also served to produce the first carbonate rocks. Nowadays, carbonates do not precipitate readily in water with a high dissolved concentration of carbon dioxide, but only when the concentration of dissolved carbon dioxide falls low, decreasing the acidity of the water. It was largely in the immediate neighbourhood of algal colonies, where the carbon-dioxide content of the water had been reduced by the demands of the algae, that carbonates could precipitate in appreciable quantity. Carbonaceous material tended to form around algal colonies. Indeed, the colonies of algae used the carbonaceous material as platforms on which they replicated themselves. Layer after layer of the material was precipitated, building into structures known as stromatolites.

As the atmospheric oxygen was produced, it was exactly sufficient to burn up all the living and fossil biomaterial. But, once some oxygen had been used to change ferrous to ferric iron, the remaining atmospheric oxygen was necessarily insufficient to burn up all the biomaterial. If geological events had caused too much fossil biomaterial to be exposed, so that the atmosphere was incapable of oxidizing it, all the oxygen would have been quickly consumed, since biomaterial burns very easily.

If this had happened, the atmosphere would have been returned to its Archean state. Yet, if the excess unburned biomaterial were again submerged, the continuing action of the blue-green algae would have started abstracting carbon dioxide and returning oxygen to the atmosphere once more. We have here the makings of a large-scale cycle. When geological events caused an oscillatory exposure of large quantities of biomaterial to the atmosphere – for instance, through variations of sea level – there would have been corresponding changes in the oxygen and carbon-dioxide contents of the atmosphere, with the possibility of the oxygen being run down to zero over part of each cycle.

Evidence for such oscillations can be found in the Great Lakes area of the United States. This region's industrial economy was founded in the nineteenth century on vast deposits of rich iron ore dating from the Lower Proterozoic. The deposits are in alternating layers, a layer rich in iron being followed by a layer poor in iron. The iron is ferric (or a modification of ferric) and it has evidently been precipitated from water. As explained, the ferric form begins as ferrous iron, which is soluble in water. Addition of oxygen to the atmosphere changes the dissolved ferrous iron to insoluble ferric, which is then precipitated as an iron layer. The apparently contradictory need for both oxygen and no oxygen has tormented geologists for a century or more. It is resolved neatly, however, by the oscillations described in the previous paragraph. The iron goes to the ferric form in the oxygen-rich phases of the oscillations, but it dissolves in water during the oxygen-poor phases, with the precipitation occurring in each cycle at the very stage at which oxygen returns to the atmosphere.

I suggested above that the Huronian glaciation of Figure 5 was caused by a collapse of temperature which followed an almost complete exhaustion of the atmospheric carbon dioxide. Since only a modest fraction of the carbon dioxide was then 'fixed' into carbonate rocks, the main quantity must have been stored in submerged or buried biomaterial. If subsequent geological events had caused a significant fraction of the biomaterial to become exposed to the atmosphere, oxidation would have returned carbon dioxide to the atmosphere. The amount could hardly have been trivial, given the high atmospheric pressure of carbon dioxide in the Archean. It could have amounted to an 'atmosphere' or more of carbon dioxide, quite sufficient to re-establish a heat-radiation trap and so drive up the Earth's temperature again, preventing further glaciations for another 1000 million years (i.e. until the Upper Proterozoic, as in Figure 5). The re-established atmospheric carbon dioxide could also have been sufficient to drive up the ocean's temperature again to a value too high for the evolution of complex life-forms.

Gradually but inexorably, however, the huge initial supply of carbon dioxide became fixed, partly in the accumulation of carbonate rocks and partly through permanent burial of the bio-material. By 'biomaterial' I mean the biological counterweight of the oxygen now present in the Earth's atmosphere (which amounts to about 2 tons per square metre) and also the counterweight of all the oxygen which has been used to convert ferrous to ferric iron (15 tons per square metre). It is an inescapable conclusion that somewhere buried in the Earth there is a huge quantity of carbonaceous material. Some of this material is straightforward coal, of which there is about one-hundredth of a ton per square metre of the Earth's surface. But there must be at least a thousand times more carbon that has not yet been found by drilling from the Earth's surface. The carbon must have been carried to considerable depths below the surface by crustal movements. At the high pressures which exist below the upper crust, this carbon must have been converted largely to graphite and diamond.

The existence of diamonds provides a clue to show that the above argument is correct. Diamonds are formed from graphite at high temperatures and pressures – indeed, at pressures which are not attained inside the Earth except at depths below the surface of more than 100 kilometres. The presence of diamonds is incontrovertible proof that free carbon is present at such depths. Between these depths and the surface there is ample room for the required quantity of carbon to be stored – at least 15 tons per square metre of horizontal area.

As I have explained, if the amount of carbon dioxide was excessive, the temperature rose too high to permit the existence of complex life-forms. Yet, if the amount was too small, there was a possibility that it would be run down completely to zero. An intermediate situation with a modest but stable supply was needed for all terrestrial life other than algae and bacteria. Such a supply was eventually provided by carbonate rocks and volcanoes.

Once the carbonate rocks became carried along in the movements of the crust, they contributed to the feed-stock of volcanoes, and carbon dioxide began to emerge into the atmosphere at a slow but more or less steady rate. Gone by the time of the Upper Proterozoic was the wholly inhospitable atmosphere of the Archean, and gone too were the wild atmospheric swings of the Lower Proterozoic. Once both the temperature and the carbon dioxide problems has been resolved, life was able to begin the dazzling evolutionary path it has followed over the past 600 million years. By then, the Earth's temperature had fallen to what we regard today as a 'normal' range, a range which made possible the pattern of recurring ice epochs.

6 The Cause of Ice Epochs

We have seen how the Archean ocean cooled sufficiently to permit the formation of the Earth's first ice sheets. What we must now establish is why there have been extensive periods of time when little or no ice formed anywhere on Earth, while at others the ice sheets, though they extended and retreated as individual ice ages came and went, never disappeared.

The first ice of our present glacial epoch formed on the Antarctic continent 35 to 40 million years ago. The climate then deteriorated slowly and steadily until the Antarctic ice sheet was fully formed about 20 million years ago. In order to find out why there was no ice in the Antarctic 50 million years ago and what eventually caused the ice sheets to develop there, we must consider a quite different phenomenon: the motion of the continents of the Earth.

When the first accurate maps were drawn, in the eighteenth century, it was seen that the shape of the west coast of Africa corresponded closely to the east coast of South America. Then, in the nineteenth and early twentieth centuries, as geologists came to classify the epochs in which the various forms of rock were laid down, it was found that rocks of similar ages on the west coast of Africa and the east coast of South America also fitted against each other when the coastlines were matched, as in Figure 22. Geologists then realized that, taking account of the shallow continental shelves, the United States, Greenland and western Europe could be fitted closely into the same pattern. The continental shelves are now covered by comparatively recent sediments which make it harder to trace similar patterns of older rocks from one continent to another, but a careful examination of Figure 22 shows that similar patterns extend across from Europe to Greenland and to the United States. The only areas of slight discrepancy are the thin black strips shown in Figure 22.

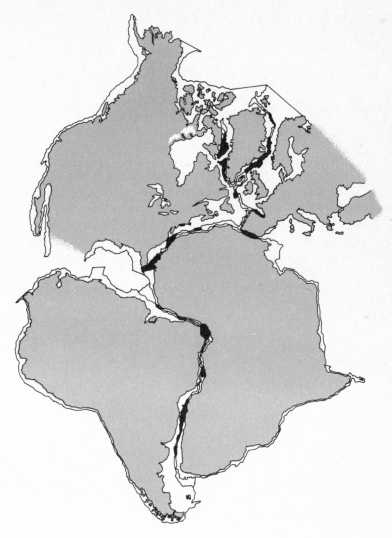

Figure 22: *The fitting of Africa, Europe, North and South America into a single supercontinent. (After P. Hurley,* Scientific American, *vol. 218, 1968.)*

(Since the continents are thought to have separated from each other as many as 200 million years ago, these areas of mismatch are really very small indeed.) The detailed fitting of rock types across the north Atlantic is shown more clearly in Figure 23.

The Baltic shield of Figure 23 is composed of rocks estimated to be older than 1700 million years – the rocks of the Russian Platform and of the African foreland were formed between 800 and 1700 million years ago, while the rocks marked in black form a clearly connected, strip-like area extending from north Greenland and north Norway through the northern areas of the British Isles to Newfoundland, and thence to the eastern coastal

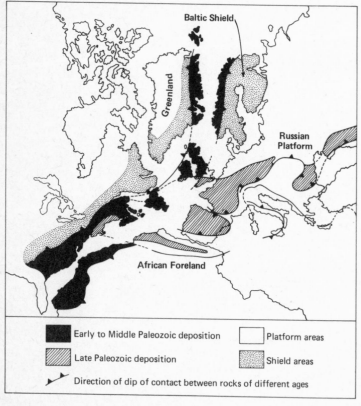

Early to Middle Paleozoic deposition

Late Paleozoic deposition

Platform areas

Shield areas

Direction of dip of contact between rocks of different ages

Figure 23: *Geological fit across the north Atlantic. (After P. Hurley,* Scientific American, *vol. 218, 1968.)*

areas of the United States and across into Morocco. The connections are obvious.

The fitting together of south-east Africa, India, Australia, Antarctica and Madagascar is shown in Figure 24 and is more equivocal. If we had to rely on Figure 24 alone, there would perhaps be grounds for the scepticism which greeted Alfred Wegener's first description of the supercontinent of Figure 22 that he called Pangea. Wegener first published his discovery in 1912. He believed that the supercontinent had broken into pieces about 200 million years ago and that the pieces had since drifted about the Earth, eventually reaching their present-day positions. The idea attracted widespread discussion and

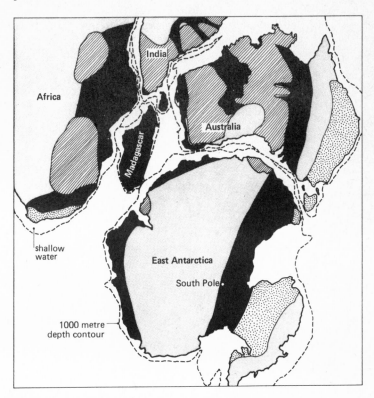

Figure 24: *Geological fit of Africa, Antarctica and Australia. (After P. Hurley,* Scientific American, *vol. 218, 1968.)*

controversy. Wegener made the mistake of attempting to defend his position against geologists and physicists who argued that no conceivable force would have been strong enough to shift the great masses of the continents apart from each other by the necessary thousands of kilometres. The forces which Wegener himself was able to imagine were clearly inadequate, and his critics lost no time in destroying his theoretical arguments so that by the 1920s the supercontinent idea had been dismissed by most scientists as a crackpot theory.

Some people, however, continued to believe in it, among them two well-known geologists, Alexander du Toit in South Africa and Arthur Holmes in Britain. The facts were not then as clearly marshalled as they are now and the sceptics continued to argue that any apparent correlations were simply coincidental. It was not until the early 1960s that E. C. Bullard put together the overwhelming evidence in a form similar to Figure 22.

Before Bullard's evidence became available, P. M. S. Blackett started a programme of magnetic measurements from the rocks of several continents. As liquid rock cools, the iron-bearing particles within it orientate themselves with respect to the Earth's magnetic field in a manner similar to small particles in the field of an ordinary magnet (see Technical Note 2). After solidification, the magnetic orientations do not change unless the rocks are reheated or buckled by violent Earth movements. Therefore, if the continents have *not* drifted around on the surface of the Earth since the rocks in question were formed, iron-bearing particles in rocks of similar ages in different continents should be orientated from one continent to another in a systematic pattern. If, however, the continents have drifted more or less randomly, the magnetic pattern would be higgledy-piggledy, because the continents would now be in positions quite different from those at which the magnetic orientations had become fixed.

As a result of work in the United States, notably by B. C. Heezen, M. Ewing and H. W. Menard, the topography of the ocean floor was becoming much better known. This work made possible the construction of such remarkable diagrams as Figure 25. The outstanding feature of the Atlantic basin is the mid-Atlantic ridge with its many transverse fracture zones. In the early 1960s, H. H. Hess and R. S. Dietz suggested that the continents had not ground their way through the ocean bed as Wegener had supposed, but that the floors of the oceans were spreading through the creation of new material in a world-wide system of rifts similar to the mid-Atlantic ridge. The idea was that the continents sit on plates and are shifted around as the plates move in the manner of a system of conveyor belts.

It occurred to F. J. Vine and D. H. Matthews that the magnetic methods of Blackett could be used to test this idea. Iron-bearing particles contained within hot material emerging from the

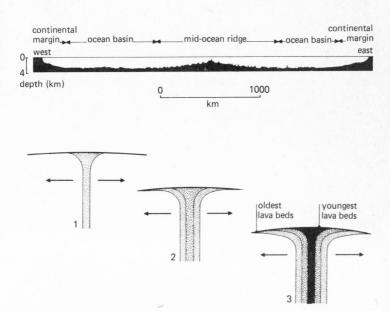

Figure 25: *Mid-ocean ridges, as exemplified by the mid-Atlantic ridge. Above, a profile across the North Atlantic from New England to the Spanish Sahara, showing the ridge with a deep basin each side. Below, a schematic diagram showing the mechanism of the ridge's growth, by repeated fissure eruptions and lateral spreading. (After: Martin H. P. Bott,* The Interior of the Earth, *Edward Arnold, 1971, and Frank Press and Raymond Siever,* Earth, *W. H. Freeman, 1974.)*

mid-Atlantic ridge would have become orientated with respect to the Earth's magnetic field as the material cooled. Since the Earth's field changes with time (Technical Note 2), the emerging material, spreading both east and west from the ridge, should contain identically changing patterns of orientation of the iron-bearing particles. This was the crucial test of the plate theory, which extensive investigations have now shown to be correct.

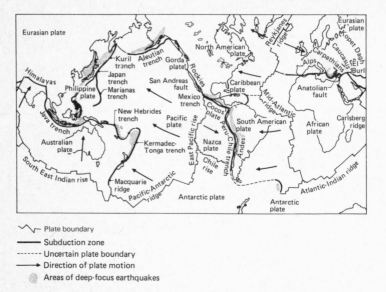

— Plate boundary
— Subduction zone
------ Uncertain plate boundary
⟶ Direction of plate motion
⟶ Areas of deep-focus earthquakes

Figure 26: *The Earth's plates. The relative motions of the plates, assuming the African plate to be stationary, are shown by the arrows. Plate boundaries are outlined by earthquake belts. Plates separate along the axes of mid-ocean ridges, slide past each other along transform faults, and collide at subduction zones. (After J. Dewey,* Scientific American, *1972.)*

Nobody today, therefore, can be in any doubt that sea-floor spreading and continental drift really do take place. It is usual that, when the apparently incredible turns out to be true, something quite unexpected has been missed. In the latter part of the nineteenth century, there was a hard-fought controversy between Lord Kelvin and geologists generally. Kelvin maintained,

The vast Byrd Glacier, one of the many great glaciers which flow
down from the central plateau of Antarctica (*US Navy*)

The Patmore Glacier, British Columbia, showing band 'ogives' of dark and light in regular spacing (*Austin Post, US Geological Survey*)

Top: Looking up from the bottom of a crevasse, with the light shining strongly through the ice (*Edward R. LaChapelle*)

The Barringer Crater, Arizona, made by the impact of a great meteorite (*US Geological Survey*)

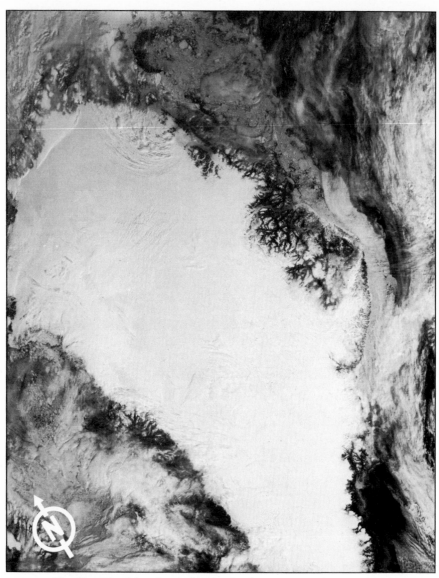

The Greenland ice sheet, showing ice-free areas on the east coast, ice floes off-shore and swirling cloud (*NOAA/NESS/EDIS*)

on what seemed to be very powerful grounds, that the Earth could be no older than about 20 million years – which geologists knew was impossible. What Kelvin missed in his argument was the phenomenon of nuclear energy, unknown and unsuspected in the nineteenth century. I suspect that something equally unexpected has been overlooked in the matter of continental drift.

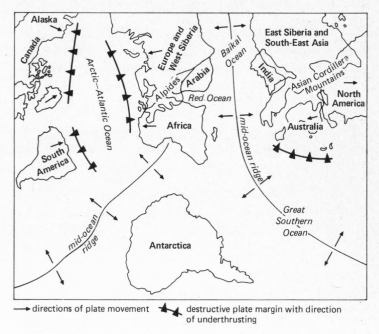

→ directions of plate movement   ⬥⬥ destructive plate margin with direction of underthrusting

Figure 27: *The predicted positions of the continents 150 million years hence. (After B. S. John (ed.),* Winters of the World, *David and Charles, 1979.)*

The main plates of the world are shown with various shadings in Figure 26. The speeds of the plate motions are known to range from 1 to 15 centimetres per year, producing shifts of the continents through distances comparable to the radius of the Earth itself (6400 kilometres) in about 150 million years. It has been estimated that, 150 million years hence, the pattern of the continents will take the form shown in Figure 27, unless the speeds of the plate motions change significantly.

As the continents drifted around in constantly varying patterns throughout the geological ages, it sometimes happened that a land mass with high ground took up a temporary position at one or other of the poles of the Earth. When this happened, ice began to lie permanently on the high ground, spreading itself gradually until an ice sheet developed.

Once an ice sheet has become fully developed, as it did in Antarctica about 20 million years ago, the normal process of calving takes place and icebergs are released into the ocean. After calving from the Greenland and Antarctic ice sheets and glaciers, icebergs can move thousands of miles in a few years. Those from Greenland have sometimes managed to get away from the Arctic by slipping south along a route close to the coast of Labrador. As the icebergs come into warmer waters, they tend to melt, but some ocean currents have enough speed to carry them into the Gulf Stream, whence they may reach to within a few hundred miles of the British Isles before finally melting.

The effect on the temperature of the ocean of the melting of icebergs is immense. It takes almost as much energy to melt ice as to raise the temperature of the resulting water to boiling point, and this energy has to come from the sea. Unlike pure water, which has its maximum density at about 4°C, salt water increases in density right down to its freezing point of about −2°C because of the salt dissolved in it. Therefore, seawater that has lost its heat in melting icebergs is very dense and plunges to the very bottom of the ocean, where it cannot be heated by the sun. All that direct sunlight can do is to heat a layer only tens of metres deep at the ocean surface. Calving of icebergs each year from the Antarctic amounts at the present time to about 1500 cubic kilometres of ice. If the sinking rate of cold dense water to the ocean floor had averaged only about 5 per cent of the calving rate, the whole volume of the ocean would have been filled with icy water in about 20 million years.

In winter, temporary sea ice forms around the Antarctic continent on the surface of the sea, freezing out of the salt solution. This leaves an excess concentration of salt in the liquid beneath the ice. Because of its low temperature and the weight of its excess salt, such water is the densest in the world, and as it is formed it too sinks to the ocean depths.

Although the Sun is powerless to heat the deep-lying, very cold water, there is an indirect method for driving heat downwards in the ocean. This method is best illustrated by using the example of the Mediterranean, where evaporation exceeds both the local rainfall and the supply of water from far-flowing rivers such as the Nile. The effect of evaporation is to raise the salt content in the remaining water, making it denser than ordinary warm seawater. Some of this very dense, warm water at the bottom of the Straits of Gibraltar manages to slip out into the Atlantic (in the opposite direction from the main surface drift from the Atlantic into the Mediterranean). Once out in the Atlantic, the escaping Mediterranean water sinks to a depth of about 500 metres because of its high salinity. The Antarctic water, however, is both highly salt and very cold. In competition with the warmer water from the Mediterranean (and of water from other areas of fast evaporation), the Antarctic water dominates at depths below 500 metres. Figure 28 shows the decline in temperature of the *bottom* water of the Pacific Ocean over the past 40 million years caused by the calving of the Antarctic ice.

The *surface* water of the oceans has also been declining in temperature over the past 30 million years, possibly because of the fresh water provided by the melting icebergs. Figure 29

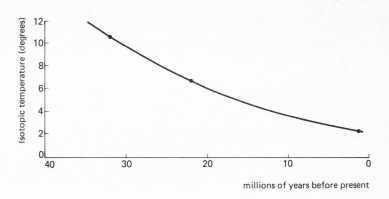

Figure 28: *The temperature of the bottom water of the Pacific Ocean has declined by 10° over the past 35 million years. (After C. Emiliani,* Scientific American, *vol. 198, 1958.)*

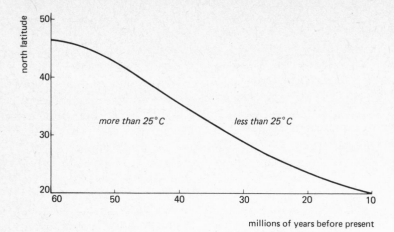

north latitude

millions of years before present

Figure 29: *The cooling coastal waters of the Pacific Ocean. The curve shows the latitude in the northern hemisphere at which the temperature in coastal waters has been 25° over the last 60 million years. The figure shows how markedly the ocean has cooled over this span of time.*

shows the latitude in the north Atlantic at which the temperature of the surface water has averaged 25°C. Plotted in the form of a graph from 60 million BP, the figure shows that, about 10 million years ago, the belt of warm surface water closed right down to the equatorial regions.

I suggest that the cooling of the oceans, caused by the arrival of the drifting Antarctic continent at the South Pole, may be said to have been the cause of our glacial epoch, an epoch in which the fine details of particular ice ages were determined by some further climatic process. If this view is correct, we might expect that other, more ancient, glacial epochs were caused by some continent similarly drifting into one of the polar regions.

The evidence of the Huronian glaciation (2300 million BP) is largely confined to the Lake Huron region of Canada (especially to the 'Gowganda Formation' of the Timiskaming subprovince of Ontario), although there is less substantive evidence from the Transvaal and north-western Australia. Some people believe that the evidence, which consists of striated pavements and boulders, does not prove that an ice age existed at that time. Although

100

the action of glaciers and ice sheets causes striated pavements and boulders, submarine mud avalanches can have similar effects. The time interval since the Huronian is now so vast and the evidence so slight that it is impossible to be certain. Nor has it yet been possible to trace the positions of the continents and their relations to the North and South Poles, so we have no means of judging whether or not the Huronian glaciation occurred contemporaneously with the filling of the deep ocean basins with ice-cold water.

Evidence of the Upper Proterozoic glaciation, however, has been found throughout the world. Some geologists have suggested that this too could have been caused by mud avalanches, but the amount of the tills that have been discovered would seem to be too vast to have been caused by anything other than glaciation. It was at one time assumed that this glaciation was confined to a single epoch around 750 million BP, and that the world-wide distribution of tills which have been discovered implied that there was then an exceedingly cold climate over the whole Earth. But, if the Upper Proterozoic glaciation spanned the enormous interval of 600 million years from 1200 million BP that many scientists now accept, there would have been ample time for two or three redistributions of the continents over the Earth's surface. Indeed, the world-wide distribution of tills from the glaciation of the Upper Proterozoic should perhaps be interpreted as evidence of continental drift occurring back in time to at least 1000 million BP. Judging from these tills, all the major continents were affected, because at some stage in the 600 million years of the Upper Proterozoic every major continent would seem to have rested in either the north or south polar regions.

The next glacial epoch began in the late Ordovician geological period, about 460 million years ago, and lasted until 410 million BP. It was largely confined to the southern hemisphere, with a vast ice sheet extending outward from the region of the South Pole, as Figure 30 shows. Figure 30 is largely based on the work of the past fifteen years, which has been excellently described by R. Fairbridge in *Winters of the World*. He vividly evokes the continuing sense of wonder as geologists uncovered evidence of an ice sheet that stretched from equatorial Africa as far as Arabia, over what are now very hot, arid deserts.

101

A close look at Figure 30 shows ice motions from land above sea level (the clear areas in the figure) into the sea itself (the cross-hatched areas). The places from which field evidence has been recovered are shown by the arrows, and these are all in the sea-covered areas. Thus, the evidence comes from the marginal regions of the late Ordovician ice sheet, which spread itself down on to extensive continental shelves, before calving off into

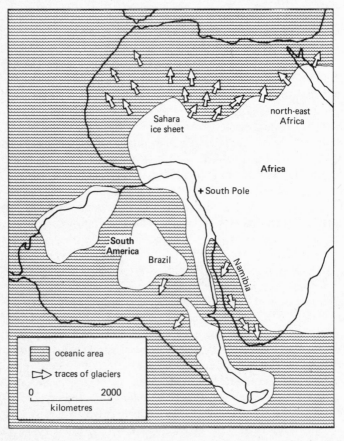

Figure 30: *Reconstruction of the Ordovician glaciation. (After R. Fairbridge in B. S. John (ed.),* Winters of the World, *David and Charles, 1979.)*

deeper waters. The deposits on these ancient continental shelves still show today in striated sandstone rocks found in the north-western African deserts.

The striations were caused by the motion of the ice. As I explained in Chapter 3, when ice rests on firm bedrock, the friction is great enough to limit basal slip to an extent consider-ably smaller than the plastic flow within the ice itself. But, when a glacier or ice sheet rests on a flat, slippery basement, such as sand or water, the ice can move more quickly at its base than at its upper surface. The plastic flow of an ice sheet resting on a smooth basement is about double that of one lying on a rough basement. The late Ordovician ice sheet rested on a rough basement as far as the land margins shown in Figure 30 and so, when it reached the smooth base of the continental shelves, the bottom ice would have shot out to great distances – just as a cherry stone does when it is squeezed out between the fingers – scoring the underlying sandy bottom with a multitude of grooves. The miracle is that these grooves can still be discovered 400 million years later by modern geologists. Dr Fairbridge remarks that, except where they are covered by loose sand, the grooves can be followed for hundreds of kilometres across the desert.

Figure 31 shows how the continents were distributed during the late Ordovician period. There was very little land in the northern hemisphere: *Australia* was the northernmost continent, while Greenland, North America, Europe, Antarctica and Siberia basked in tropical warmth. As the British Isles enjoyed an equatorial climate, there is no evidence in Britain of the late Ordovician glaciation. But the warm climate continued and left its mark there. The rich, red soil of Devon was formed about 400 million years ago, together with other deposits of Old Red Sand-stone, at a time when the British Isles, as part of the European continent, lay close to the equator.

Figure 30 shows western equatorial Africa lying at the South Pole. This position has been determined by M. McElhinny using the rock magnetism method described above, and is strong evi-dence that the arrival of a continent at one of the poles initiates an ice epoch.

About 90 million years after the last of the Ordovician ice disappeared, another glacial epoch began. This was the great

Permo-Carboniferous glaciation, which lasted for 100 million years, reaching its height about 280 million years ago. It is probably not a coincidence that 340 million BP was of immense significance for biology. It was then that the first insects appeared and the first vertebrates crawled out of the sea to try life on land. When we remember that the Cenozoic ice epoch saw the emergence of man himself, the suggestion that there is a connection between decisive steps in evolutionary biology and upheavals of the world climate becomes difficult to ignore.

Overwhelming evidence of the Permo-Carboniferous glaciation comes from Brazil, South Africa, India, Australia and the Antarctic. The vast area involved is shown in Figure 32. The evidence from Antarctica has been found only recently, but the existence of polished and striated bedrock and of thick deposits of glacial till was discovered in India as early as 1859. One can well imagine the astonishment of geologists when they first came on the Indian striated pavements and tills and wondered how there could ever have been glaciers in a place as warm as India.

The Indian glaciers had been big enough and lasted long enough to deposit tills of enormous depth – as much as 250

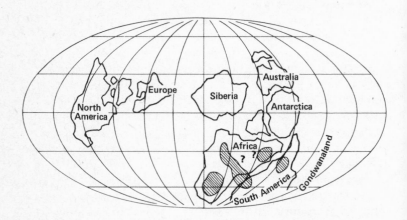

Figure 31: *The distribution of the continents in late Ordovician times. Possible ice sheets are marked by the hatched areas. (After B. S. John (ed.),* Winters of the World, *David and Charles, 1979.)*

metres in some places. Still greater depths of till have since been found elsewhere – up to 1000 metres in Antarctica and South Africa, and up to a fantastic depth of 1600 metres in Brazil. These huge, unsorted deposits of stones and grit attest to the long timespan of the Permo-Carboniferous.

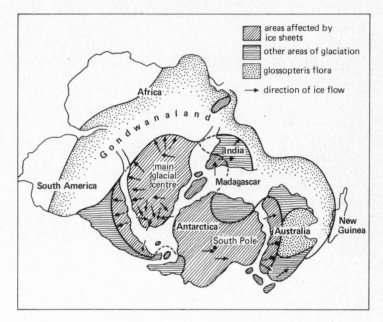

Figure 32: *The Permo-Carboniferous glaciation. South America and South Africa appear to have been glaciated before Antarctica and Australia. (After B. S. John (ed.),* Winters of the World, *David and Charles, 1979.)*

If individual ice ages came and went every 100,000 years or so, as they have done in the present Cenozoic period, there must have been 500–1000 individual ice ages in the great Permo-Carboniferous epoch. Nothing like this number of individual episodes has actually been counted, although evidence for up to nearly twenty subperiods of intense glaciation has been found. It is unlikely that the whole of the vast ice sheet drawn in Figure 32 melted away during the warmer periods, any more than the ice

105

sheets of Greenland and Antarctica have disappeared in our present warm episode. Rather, it would be a case of a retreat of the ice sheet at its margin, a shrinkage of its area and a decrease in its thickness. The great depths of till are distributed around the periphery of the ice sheet, just where the melting would have been most frequent. The enormous extent of the till gives unequivocal evidence of many disappearances of the marginal ice. On each occasion the ice, as it melted, deposited a further layer of debris into the till.

The point marked 'South Pole' in Figure 32 is of course the present-day pole. Over a timespan as long as 50 million years (from 310 million BP to 260 million BP, i.e. the central, most intense, part of the Permo-Carboniferous glaciation), the position of the continents must have changed appreciably. There was, for example, a period when south-west Africa was at the pole. Such variations must have caused changes in the ice sheet at its outer margin and in the relative thickness of the ice at different places within the sheet. About 280 million BP, the distribution of continents took the form given in Figure 33. It shows four unusual names: Angaraland, for the northern continent; Laurasia, for the equatorial lands; Tethys, for the ocean; and Gondwanaland, for the huge southern glaciated continent.

After 280 million BP, Laurasia turned around in a clockwise

Figure 33: *The distribution of the continents in late Carboniferous times. The position of ice sheets is marked by the hatched areas. (After B. S. John (ed.),* Winters of the World, *David and Charles, 1979.)*

direction to bring its leading edge up against the south-west coast of Angaraland. The resulting pressure lifted a range of mountains which today we call the Urals. The frictional heating generated by this fusion of Laurasia and Angaraland melted rocks, causing volcanoes to burst out and ores to be deposited. Among the ores were thin veins of platinum and other precious metals. Subsequently, however, the continents separated again, but not at the junction along the Ural Mountains. When this separation took place, Siberia gained territory from North America – the territory of Russia itself.

Land masses were also pressed to the south by a clockwise turning of Laurasia, creating a log jam there like a nearly immovable position on a chess board. There were widespread fusions which eventually formed the pieces into a single continent (the Pangea of Alfred Wegener) stretching in a vast arc over the Earth from the North to the South Pole.

The pressures generated when land masses were fused lifted mountains which we call Hercynian. The results of this fusing and buckling of the land are still with us today, in the Appalachian mountain chain, the lanes and fields of the counties of Devon and Cornwall, and the hills of south-west Ireland from Kerry to Cork. These are the bones of Hercynian mountains. They are there too in the Massif Central of France, the Hartz and Vosges mountains, the Black Forest of Germany, and also in the beautiful country of southern Poland and Bohemia.

The emotional influence of these landscapes on man has been captured in Smetana's *Ma Vlast*. Is it because the people of Bohemia, southern Poland, southern Germany, central France, south-west England and Ireland, and American Appalachia all share in a common geological heritage that they tend to feel Smetana's music belongs not only to Czechoslovakia but to them too?

The Permo-Carboniferous glaciation served us well in quite another way, for it was then that most of the coal, on which our modern society must surely depend in its present energy crisis, was laid down. If you rotate Laurasia clockwise in Figure 33 until it abuts Angaraland, so forming the Ural Mountains, and then recall where the world's greatest deposits of coal have been found, a very curious and interesting picture emerges.

107

By far the greatest deposits are in the far north, and they decrease in a band that runs north-east to south-west from Siberia through Russia, Poland and the United States, thence declining into western Europe. It was evidently hardy plants living in periglacial regions that were turned into coal. Just the same picture is seen in the southern hemisphere, for it is in the glaciated, or near-glaciated, lands of South Africa, India and Australia that the southern coal measures have been found.[1]

The amount of ice on the land was so huge that the melting of only a small fraction of it would have raised sea level by tens of metres – sufficient to inundate forests that had grown close to the ocean. The Permo-Carboniferous forests should not be thought of as being anything like the forests of today. It was not for a further 150 million years that flowering plants appeared on the Earth. (The term 'flowering plant' signifies more than plants with coloured flowers: it encompasses all plants that encase their seeds in fruits, including deciduous trees and grasses.) Even the conifers had barely made an appearance by the end of the Permian period. Land plants were predominantly mosses and ferns, and it was from ferns that Permo-Carboniferous coal is very largely derived.

For the vegetable matter not to have been broken down by bacteria, fungi and other protozoa, the raising of sea level must have occurred rapidly. Biological degradation would occur much quicker in the tropics than in the cool regions at high latitudes, and so it seems reasonable to suppose that biodegradation occurred in the tropics more rapidly than the raising of sea level. In periglacial regions, however, sea level must have risen more rapidly than biodegradation could take place. The concentration of coal in such regions can therefore be explained in terms of a competition in speed between biodegradation and the raising of sea level. Degradation would occur in a year or two under warm conditions, and perhaps in a century under cool conditions. Sea level must, therefore, have risen by some tens of metres in only a few decades. There was clearly the same rapid melting of ice in

[1] South America has not yet been thoroughly explored. One would suspect from these considerations that considerable quantities of coal will eventually be found there.

the Permo-Carboniferous that there has been at the end of the ice ages of the recent 2 million years.

Sea level must have continued to rise after the vegetation was inundated, because there are marine deposits that overlie the Permo-Carboniferous coal measures. Shales and limestones are found, the latter rich in fossils of marine animals. On the other hand, limestones found *below* the coal measures contain fossils of freshwater animals, proving that the coal marked a sudden change of sea level.

Coal is often found in several layers, with each layer 1 or 2 feet thick in the economically viable deposits. The coal layers are separated by thicker deposits of the shales and limestones. This multi-layering was caused by repeated changes in sea level, corresponding to oscillations of the amount of ice piled on the land. In this sense, there were evidently repeated ice ages within the ice epoch, just as there have been in the present Cenozoic epoch.

Having established that there was a land mass at one or other pole during each glacial epoch, causing the loss of a huge amount of energy from the sea, we must now find out why there have been repeated ice ages within those epochs, alternating with relatively warm periods, such as the one we are temporarily enjoying today.

7 The Earth as a Heat Engine

So far, we have looked mostly at bits of the Earth's surface – at boulders, glaciers, mountains and seas. The time has come, however, to think of the Earth as a whole. Pictures of the Earth taken from the Moon show a predominantly blue colour, which is caused by the scattering of light by air molecules in the Earth's atmosphere. Part of the blue light incident from the Sun is scattered forward on to the Earth, which is why the sky normally appears blue to us, and part is scattered back into space, which is why the Earth appears blue to a person on the Moon. It is estimated that this scattering by air molecules has the effect of reflecting back into space about 9 per cent of the energy of all the sunlight incident on the Earth.

The strongest reflection of sunlight, about 90 per cent, comes from new snow; clouds have an average reflectivity of about 50 per cent; ice and deserts about 35 per cent. Land areas are generally a good deal lower in reflectivity, however – usually 10 to 20 per cent, depending on the nature of the vegetation that the land area in question happens to carry. The ocean, which covers 71 per cent of the Earth's surface, is least reflective of all – only about 3 per cent (which is why the ocean appears dark in pictures of the Earth taken from the Moon or from artificial satellites). When all sources of reflection are added together, our planet is found to turn back into space some 36 per cent of the solar radiation falling upon it. It follows that the Earth has available only 64 per cent of the energy of incident sunlight.

It might seem very difficult to calculate the precise degree of heating produced by the Sun, since the Earth is an extremely complex body. Indeed, a complete calculation including every detail would be a daunting affair, for which it would be necessary to use a battery of the largest and fastest computers. One might spend a lifetime and still not complete the problem. It is therefore surprising that the simplest conceivable approximate calculation gives an answer that comes very close to the actual measured average temperature of the Earth.

Instead of the complex Earth, consider a small solid spherical body with the following elementary properties: (1) the body absorbs all solar radiation incident upon it; (2) the body is a good radiator of heat at all wavelengths; (3) the body manages to average its temperature by a rapid transfer of heat from one part of itself to another (like conduction by a piece of metal).

Now suppose the object moves around the Sun in the same orbit as the Earth – that is to say, at an average distance from the Sun of 150 million kilometres. What temperature would we expect it to have? The answer is about 280 Kelvin (7°C), a result close to the actual average temperature of the Earth, which is about 287 Kelvin (14°C).

Let me add a few words about temperature measured in degrees Centigrade and in degrees Kelvin, and about the meaning of temperature itself. If one takes the temperature *difference* between two bodies, the answer is just the same whether one uses the Centigrade scale or the Kelvin, but the two scales start from different zeros. Pure ice placed under a pressure equal to the weight of a 76 centimetre column of mercury (a 'standard atmosphere') melts at 0°C. On the other hand, the zero temperature of the Kelvin scale, 0K, corresponds to the condition of a body which has no available internal heat at all. Temperatures on the Kelvin scale are a measure of the internal heat of a body – any body – which gives to the Kelvin scale an absolute quality that is lacking for the Centigrade scale, since the latter is based on a special property of one particular substance – ice. (If the melting of frozen ammonia instead of ice had been used to determine the Centigrade scale, the zero point for the scale would have been quite different.) It is easy to pass from Kelvin to Centigrade and *vice versa*. Add 273 to a Centigrade temperature and you have the Kelvin temperature. Subtract 273 from a Kelvin temperature and you then have the corresponding Centigrade temperature.[1] Because of its absolute quality, it is the temperature in the Kelvin scale that must be used if we are to estimate the discrepancy

[1] More precisely, the number to be added or subtracted is 273.15.

111

between the result calculated for the imaginary body and the actual average temperature of the Earth. The error of the calculation was less than 3 per cent.

To upgrade our simple calculation we must make the imaginary body more realistic. Suppose it reflects 36 per cent of the sunlight incident upon it, as the Earth does, but otherwise keeps to the same conditions as before. The calculated temperature is now about 250K, giving a 13 per cent error. This worsening of the error suggests that the Earth itself must have some feature which compensates for the energy loss caused by the reflection of sunlight.

The compensating feature is that the Earth is not a completely efficient heat radiator. The atmospheric gases of the Earth generate two radiation traps, one due to water vapour, the other to carbon dioxide. The carbon-dioxide trap is very strongly absorptive over a wavelength range from 14 micrometres to 16.5 micrometres (1 micrometre is a millionth part of a metre), and the water-vapour trap blocks the escape of heat with wavelengths longer than 20 micrometres.

When the wavelength falls into one or other of these blocked ranges, scarcely any radiant heat generated at ground level succeeds in penetrating the trap. There is essentially no escape through the trap into the higher atmosphere and thence into space. Blocked radiation is either re-emitted (after being absorbed in the trap) downward to the ground immediately, or absorbed again by the gas generating the trap, in which case there is a further re-emission. Sooner or later, perhaps after many absorptions and re-emissions, the trapped radiation reaches the ground, where it is removed by ground absorption, so returning its energy to where it started. Radiation with wavelengths in the carbon-dioxide and water-vapour traps simply goes round in closed cycles, and very little escapes into space in each cycle.

Heat radiation from the surface of the Earth occurs at wavelengths both inside the traps and outside them. The fraction of the energy of the radiation inside the traps is about 42 per cent, with the carbon dioxide blocking about 15 per cent of the heat energy and the water vapour the remaining 27 per cent. The effect of the traps is, therefore, to reduce the radiating efficiency of the surface of the Earth to 58 per cent of the imaginary body used in the simple calculations described above. The surface of the Earth cannot cool itself as efficiently as the imaginary body would be able to do.

The inability of the surface of the Earth to cool itself is compounded by partial trapping at wavelengths outside the ranges I have just considered. As well as the more or less complete trap for wavelengths longer than 20 micrometres, water vapour generates a partial trap for wavelengths between 16.5 micrometres and 20 micrometres, while carbon dioxide produces a partial trap between 13 micrometres and 14 micrometres. A partial trap is one from which an appreciable fraction of the radiation does get away into space, although a considerable fraction also returns to the ground, where it is reabsorbed.

There is additional partial trapping at *all* wavelengths if there happens to be an appreciable quantity of water droplets or ice crystals suspended in the atmosphere. This further trapping is most marked when cloud is heavy. It is this effect which in summer produces a hot, stifling feeling before a thunderstorm, and why we speak of heavy rainfall 'clearing the air'. The 'clearing' consists of the removal of a radiation trap previously caused by water droplets.

The partial traps are usually estimated to increase the blocking of heat energy from the 42 per cent given above to approximately 63 per cent. This calculation cannot be exact because the distribution of cloud from one part of the Earth to another is very irregular. Nevertheless, as a general statement, we can say that, of the heat energy radiated from the Earth's surface, about one-third escapes into space and the other two-thirds are reabsorbed.

This is not quite the complete picture. There is one remaining item still to be included. We saw in Chapter 5 that heat radiation traps can be circumvented. Evaporation of water at the ocean surface cools the ocean. The resulting water vapour rises in air columns to great heights, carrying latent energy with it. The latent energy is released into heat when the vapour condenses into ice crystals, which it generally does at heights above the radiation traps. The latent heat is then radiated freely into space. This bypassing of the traps is by no means complete, but it has the effect of reducing their blocking ability from the previous 63 per cent back to about 40 per cent.

The Earth's surface receives 36 per cent less solar energy per square metre because of reflection, but this loss is slightly over-compensated by a 40 per cent reduction in the rate of heat radiation. The small energy profit of 4 per cent raises the temperature of 280K obtained for the imaginary body to about 283K (10°C) for the average temperature of the Earth.

Figure 34 shows world temperatures for January and July. (These maps are drawn on the familiar Mercator projection, which is well known to exaggerate quite grossly the apparent areas of the high-latitude zones.) Considering that half of the total surface of the Earth lies in an equatorial belt between latitudes 30° south and 30° north, it is clear that the average temperature is in fact about 14°C. Indeed, the average annual temperature in the equatorial zone is about 25°C, in mid-latitudes about 10°C and in high latitudes about −10°C, so that, by weighting the areas of these three regions in the ratios 3:2:1, one obtains 14°C (287K) for the average temperature. With an ambitious use of computers, it would doubtless be possible to close the gap between observation (287K) and calculation (283K) still further. But, in doing so, one would be likely to lose sight of the really salient points, which are that the heat-radiation traps in the lower atmosphere, known as the 'greenhouse effect', compensate very closely for the loss of solar radiation due to the reflectivity of the Earth; and that about one-fifth of the incident solar energy goes into the evaporation of water vapour. The latent heat of recombination of the water vapour is released for the most part high in the atmosphere.

We are now coming close to the crux of the ice-age problem. Either we have to argue that a planet with an average temperature some 15° above the freezing point of water can spontaneously develop ice ages without any substantive cause, or a crucial datum used in the above calculations must be changed. I do not myself believe that it is at all plausible to suggest the first of these possibilities, because the Earth is such an efficient heat engine.

An engine transfers heat from a hotter region (the cylinders of a car for example) to a cooler region (the exhaust), and in so doing generates mechanical motion. The Earth transfers heat from the equatorial ocean to the polar regions, increasing the

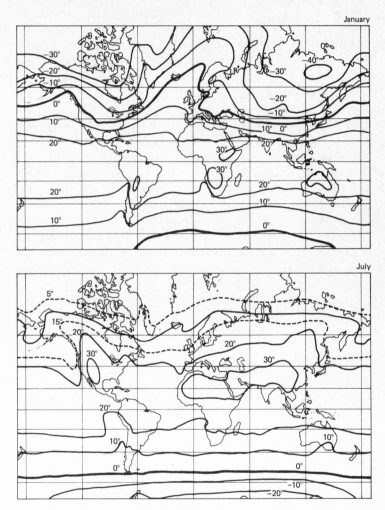

Figure 34: *Present-day world temperatures for January and July. (After J. Blüthgen,* Allgemeine Klimatgeographie, *de Gruyter, 1966.)*

115

energy available there by 80 per cent. Although the importance of oceanic heat in the amelioration of climate at high latitudes is well known to oceanologists and meteorologists, there has been curiously little mention of it in connection with the ice ages. One reason for this strange omission may be that the mode of transfer of heat is subtly concealed. Heat is not transferred through the actual movement of ocean water; it is not a case of the surface currents shown in Figure 35 carrying energy directly in a rush from equator to pole. The currents do carry some heat, but the major effect comes from water vapour in the wind. Water vapour, which yields up energy as it condenses into rain and snow, is the energy carrier and the winds are the energy distributor – a much more efficient distributor than the comparatively slow-moving ocean currents. The real importance of the ocean currents is that they move bodies of warm water some way, but by no means the whole way, towards the pole, as for example the Gulf Stream, as shown in Figure 35. Vapour evaporates from the warm water, and wind then broadcasts the energy latent in the vapour over great areas which the ocean currents themselves cannot reach.

People in the British Isles have a habit of complaining about their weather. Because of the cyclonic storms which sweep in persistently from the Atlantic, British weather is never fine and clear for very long, luckily! Without those storms, very little warmth would reach the polar regions and the glaciers would be on their way back.

Cyclonic storms are miraculous happenings. According to elementary atmospheric physics they should not happen at all. When one pushes a commonplace object, it tends to move in the direction in which it has been pushed. This sensible behaviour shows itself most clearly when one gives a flat stone a push along the surface of a smoothly frozen lake. When a big pocket of air is pushed, however, it moves away more or less at right angles to the push, sideways to the right in the northern hemisphere and to the left in the southern hemisphere. This contrary behaviour arises because the push acting on a pocket of air is not the only force acting on it. Once it starts to move in the direction of the push, a continuing force caused by the Earth's gravity then forces the pocket of air into a sideways motion.

Cyclonic storms show this effect quite clearly. The storm centre is a region of low atmospheric pressure, and so air within the system tends to be squeezed towards the centre by surrounding air. Instead of simply flowing inwards and so equalizing the interior low pressure to the high pressure outside, the air moves sideways in a circular motion around the centre of low pressure, anticlockwise in the northern hemisphere and clockwise in the southern hemisphere. A similar effect occurs when there is a centre of abnormally high pressure – an anticyclone. In this case,

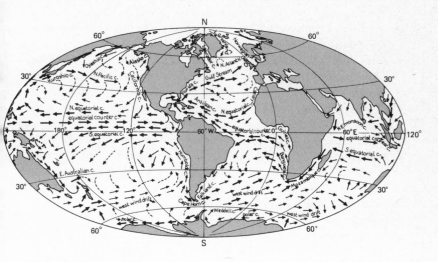

Figure 35: *Main currents in the upper waters of the world ocean. (After John G. Harvey, Atmosphere and Ocean, Artemis, 1976.)*

the obvious direction for the air to move would be outward, to relieve the central high pressure, but instead there is again a circular motion around the centre, clockwise for an anticyclone in the northern hemisphere and anticlockwise in the south. The root cause of these physical examples of Murphy's law is the relation of gravity to the spin of the Earth, the force that

117

generates the contrary behaviour being known as the 'Coriolis force'.

Unless one supplies external work to a physical system (as with a refrigerator), heat inevitably flows from the hotter parts to the cooler. This implies that heat would inevitably be transferred from the Earth's tropical regions to the poles. If the Earth were without spin, warm air would flow directly to the polar regions in some very general planet-wide atmospheric circulation, but the Coriolis force caused by the actual spin prevents that. No sooner does air have a tendency to move poleward than it is forced sideways, to the right in the north and to the left in the south, which means that the air moves west to east in both hemispheres.

Imagine a plane drawn through the Earth's axis of spin. The plane determines a pole-to-pole section of the atmosphere. Figure 36 shows two such sections in rectangular form – one for January, the other for July. Imagine westerly winds (i.e. winds that are moving from west to east) flowing downward into the paper, and easterly winds flowing upward out of the paper. Contours of equal wind speed in metres per second are shown in Figure 36, with positive speeds indicating westerlies and negative speeds indicating easterlies. Only in the tropics are there easterlies – the east-to-west trade winds used by the old sailing ships, ships laden with spices and other exotic cargoes – and the 'great' westerlies dominate the wind system of the Earth. On the equator itself there is a dead zone known as the doldrums – a term more widely used these days in social affairs than in meteorology.

If one looks at the numbers attached to the contours of Figure 36, it becomes clear that wind speeds increase towards little tubes (one in each hemisphere) at heights of about 12 kilometres in the atmosphere (about 35,000 feet). The sections of these tubes, marked in Figure 36 with a W at their centres, are the jet streams. In each hemisphere the jet stream moves towards the equator in winter. There is also a tendency in winter for speeds to increase to their maximum, about 40 metres per second (about 90 miles per hour). Modern commercial aircraft fly at just the height of the jet streams. Flying from London to Los Angeles the aeroplane has to battle with the westerlies, but in returning it rides with the wind. That is the reason why the east-to-west flight takes two or three hours longer than the return journey.

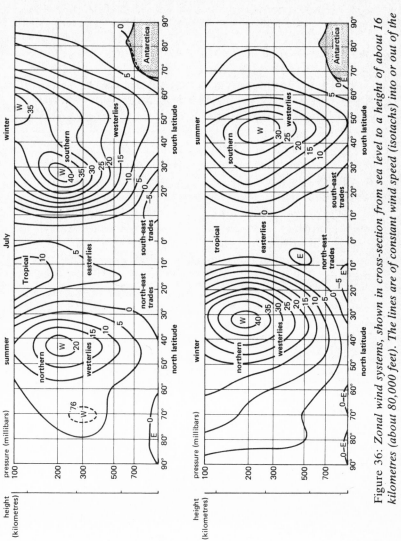

Figure 36: *Zonal wind systems, shown in cross-section from sea level to a height of about 16 kilometres (about 80,000 feet). The lines are of constant wind speed (isotachs) into or out of the plane of the cross-section (i.e. west is positive, east negative). (After H. L. Crutcher, R. L. Jenne, H. van Loon and J. J. Taljaard, Climate of the The Upper Air, part 1, Southern Hemisphere, US Naval Weather Service Command NAVAIR, 50-1C-55.)*

All this is elementary – the great westerlies frustrate the tendency for air to transfer heat directly from the warm tropics to the cold poles of the Earth – but let us come now to the cyclonic storms. Within themselves, in the circulating patterns of cyclonic storms, the same odd behaviour still holds: instead of 'filling in', the air within cyclones swirls anticlockwise in the north and clockwise in the south. But, with the storm tracks of the world, besides the usual west-to-east motion, there is also a clear tendency for the direction to be towards the poles. Storm tracks somehow evade the usual cussedness of the general motion of the air. They disobey the simple rules, so managing to transfer heat from the warmer equatorial belt towards the poles.

Unlike the meteorologists, I cannot claim to understand the ability of storms to move polewards. For me, it is one of many examples of the boundless ingenuity of nature. Physical systems left to themselves tend to even up the distribution of their internal energy; heat passes from a hotter part of the system to a cooler part. But there are no broad laws or principles of physics to suggest how quickly such levelling-up takes place. Each case has to be dealt with on its merits, with all possible processes calculated in full detail.

For a system of great complexity like the Earth, the calculations are impossible to make accurately, because the sheer amount of arithmetical labour involved goes far beyond the capacity of even the largest available computer. All one can do is to simplify the work by omitting the complications that seem to be irrelevant, hoping that one's guesses will turn out to be correct.

Levelling-up of the energy distribution within a system is often referred to as 'degradation', or more technically as an increase of the 'entropy', entropy being a mathematical quantity that measures the amount of the levelling-up. Experience shows that estimates of the rate at which systems degrade themselves are more often wrong than any other kind of calculation in physics. Experience also shows that we are very much more likely to underestimate the degradation rate than to overestimate it. And this is just what I should have done if I had tried to calculate mathematically how fast heat is transferred from low to high latitudes of the Earth. I should not have anticipated the cyclonic storms. Even though I have experienced such storms over a lifetime, I still do not really understand how they happen. Sliding of air from greater to lower heights has something to do with it, but the precise details are obscure. It is as if nature, forbidden by its own laws from a simple solution to the problem of transferring heat from equator to pole, gives the cards a quick shuffle and then flicks out a joker from the bottom of the pack. The shuffling goes a bit too fast for my eyes to follow. But it does happen and that is really all we need to know.

Let me summarize the position we have now reached. In spite of the very cold water that fills the deep ocean, the heat content of the top 500 metres plays a crucial role in preventing the immediate development of polar ice sheets. Regions in high northern latitudes obtain a third or more of their heat from the oceans through the agency of storms with tracks that generally run from south-west to north-east, carrying water vapour with them. The heat is released when the water vapour eventually turns into rain or snow, a process that usually takes place in cloud formations, high in the air. Heat released into the air is remarkably effective at removing snow and ice from the land (for example, the Föhn wind), whereas direct sunlight is ineffective. Thus, the heat released into the air by the condensation of water vapour is especially important in the prevention of ice ages.

The ability of the ocean to develop a vigorous storm pattern depends on its own heat supply. If that supply declines, the ability to drive storms up into the high-latitude regions must also decline, with the inevitable result that permanent ice cover develops there, particularly in the high-latitude regions of the northern hemisphere where the critical land areas of North America, Europe and Asia are located.

It has been well known for nearly two centuries that the greater the ratio of the temperature of a heat source to that of its heat sink, the greater the mechanical efficiency of the engine. It was by using a steam condenser to lower the temperature of the heat sink that James Watt improved the efficiency of the first steam engine of Thomas Newcomen, and it was by using superheated steam to raise the temperature of the heat source that Richard Trevithick and Oliver Evans improved its efficiency still more. It was the use of high-temperature steam that permitted the development of George Stephenson's locomotive, making a first long stride towards the tremendous technological developments of the present century.

If an ice age were to develop without there being any change in the basic features which led us to a substantially correct calculation of the Earth's present average temperature, there would be no reason for any decline of temperature in the equatorial regions. The spread of ice sheets into mid-latitudes would therefore increase the efficiency of the Earth's heat engine by increasing the contrast between its hot and cool regions (the heat source and the heat sink). There would be increased precipitation and more efficient transfer of heat towards the glaciated regions, inevitably melting the ice. The inevitable increase in precipitation and wind, moreover, contradicts the evidence of earlier chapters, which showed that ice ages of the past have been rather dry and windless. If the heat transfer were somehow mysteriously reduced, the tropics would retain more heat than they have now, making the development of ice domes on tropical mountains most unlikely.

It is evident, therefore, that some datum in the calculations must be changed if we are to understand the climatic condition that led to the ice ages. If one troubles to think back over the argument, it is clear that, for an ice age to develop, either or both of the radiation traps in the atmosphere must be reduced, or the reflectivity of the Earth must be increased.

Let us look first at the radiation traps. With an average temperature of 14°C (with 25°C occurring in the tropics), the evaporation rate of water into the atmosphere must always be very much greater than is necessary to maintain the water-vapour trap. Only the carbon-dioxide trap is therefore relevant to this discussion. As I explained earlier, the carbon-dioxide trap is highly effective over a wavelength range from 14 micrometres to 16.5 micrometres. By blocking the escape of heat radiation with wavelengths in this range, the carbon dioxide reduces the radiating efficiency of the Earth by 15 per cent.

If carbon dioxide were entirely removed from the atmosphere, the radiating efficiency of the Earth's surface would rise from 60 per cent to 75 per cent. Keeping the same reflectivity as before (36 per cent), it is easy to calculate that the average temperature of the Earth would fall to 270K ($-3°C$). At this significantly lowered temperature, it is certainly possible that an ice age would occur.

The idea that a removal of the carbon-dioxide trap caused the ice ages was suggested more than half a century ago by the Swedish chemist Svante Arrhenius. It was an idea altogether more plausible than the theory discussed in Chapter 4. The Arrhenius theory hit at the root of the problem by identifying an effect of the required magnitude, whereas the effects described by Milankovitch do not change the average temperature of the Earth at all. Ironically, it has been the zero-effect theory of Milankovitch which has received more notice over the years than the astute proposal of Arrhenius, for a reason that I shall now explain.

The efficiency of the carbon-dioxide trap is insensitive to the amount of carbon-dioxide in the atmosphere: increasing the amount five-fold would scarcely change the trap, in spite of the stories that are currently being circulated by environmentalists. Only if the amount of carbon dioxide were enormously increased (to the extent described in Chapter 5) would the trap widen its influence significantly. The trap would not contract very much either, unless the amount of atmospheric carbon ran down almost completely – a condition that would produce a cata-strophic reduction in the growth of vegetable material, leading in turn to extinctions of animals of all kinds, since animals live by eating vegetation or by eating other animals that eat vegetation. The flaw in the Arrhenius theory is that there is no evidence of such world-wide biological effects occurring at the onset of an ice age – evidence that would have been found if it existed.

Neither the water-vapour nor the carbon-dioxide traps can be changed appreciably, and so we are left with the possibility of a change in the reflectivity of the Earth. At first sight, this too seems unpromising, because the Earth's cloudiness appears to be inflexibly adjusted to its average temperature, but in the next chapter I shall show that the Earth's reflective capacity is actu-ally poised on a knife edge. The knife edge turns out to be so sharp that the balance could swing at any moment, sending the Earth into an ice-age condition in only a few years.

8 How the Engine Stops

The temperature in a rising column of air falls until water vapour in the air becomes saturated – that is to say, the vapour tends to condense into liquid droplets or into ice crystals. The temperature at which saturation occurs depends on the moisture content of the air. Humid air becomes saturated low in the atmosphere, at temperatures well above freezing point. Condensation then occurs into a swarm of tiny water droplets (it would take 1000 such droplets placed consecutively in a line to cover a distance of 1 centimetre). The water droplets scatter light, behaving as a 'fog' that reduces visibility. The 'fog' may be literally a fog, or a cloud, or simply the whitish haze common to the British Isles in summer.

Drier air has to rise higher before its water vapour becomes saturated. When the air is particularly dry, the temperature at which saturation occurs falls below freezing point and then the vapour tends to form ice crystals. There is hardly any lower atmospheric air so dry that saturation does not occur when it has risen to heights at which the temperature falls to between $-10°C$ and $-20°C$. The vapour does not automatically condense into ice crystals, however. Nor do water droplets freeze automatically into ice at such temperatures. If undisturbed, they remain liquid, becoming 'supercooled'.

In the higher air, there is a comparatively small number of particles of a different kind, however, known as condensation nuclei. The water vapour supplied by the many supercooled droplets passes over to these condensation nuclei to form ice crystals, which, because of their much smaller number, grow to comparatively large sizes. As they become bigger, the ice crystals are better able to fall under gravity, but they may encounter rising air which lifts them upwards again. In their general swirling motion, the ice particles collide with the tiny supercooled water droplets, which immediately freeze, adhering as they do so to the larger particles, which thereby grow heavier still. Eventually, the heavy particles break loose from the cloud in which they have hitherto been kept swirling. They fall lower and become warmer as they near the ground. If the ice melts before reaching the ground, it falls as rain. In winter, however, the lower air may be too cool to melt the falling ice particles, which then fall as snow. (The beautifully complex form of snow crystals arises from the joining of the many smaller particles which came together in their formation.) Occasionally, even on a warm summer day, the falling ice particles fail to melt before they reach the ground. This happens when updrafts of air in the parent cloud are so strong that the particles have to gain a quite unusually large size before gravity can cause them to fall. Being exceptionally large, sometimes as large as a golf ball, they then come down quickly through the warm lower air without being melted. This is what we call summer hail.

The swirling together of large ice crystals and tiny supercooled water droplets produces the electric charge separations that cause lightning; the more violent the turbulence, the greater the electrical effects and the larger the raindrops or hailstones that fall to the ground. Big clouds with strong turbulence, heavy precipitation and lightning, with its accompanying thunder, have therefore a natural tendency to occur in association with each other.

Everything in this condensation process turns on the condensation nuclei already present in comparatively small numbers in the upper regions of cloud – regions where the temperature is about $-20°C$. Such particles consist of ice which has formed around what are called 'freezing nuclei', which are often tiny grains of salt that have been sprayed up into the atmosphere from the sea. Water vapour at $-20°C$ will not form ice crystals by itself; freezing nuclei are needed. If there happens to be a negligible number of freezing nuclei – a possible condition – the temperature of the water vapour as it rises in the air falls still further, towards the outer boundary temperature. The outer boundary temperature is that at the top of the whole water-vapour distribution of the atmosphere. I will explain in a moment how the 'top' is to be defined, and how the boundary temperature is to be calculated. The point I wish to make here is that, if the boundary temperature is above $-40°C$, nothing especially dramatic happens. Without freezing nuclei, the sky remains clear – there is no cirrus cloud caused by ice crystal formation – but, if the boundary temperature falls below $-40°C$, small droplets of supercooled liquid change spontaneously into ice crystals, themselves becoming freezing nuclei around which more ice condenses. These are the ice particles known to Antarctic explorers as diamond dust.

Compared to the comparatively simple condensation of water droplets at higher temperatures, the conditions for ice crystals are quite complex, which is why cloud formations at low temperatures in the high atmosphere show so many varied and beautiful forms.

Ice crystals reflect and refract light just like diamonds, with the refraction producing a separation of the light into its constituent colours. When many ice crystals are randomly orientated, their effects become jumbled together, so that we see nothing but a dull, whitish haze. But, if the crystals become systematically orientated, their optical effects also become systematic. The simplest effect occurs when needle-shaped crystals are systematically orientated by gravity. This particular orientation causes the well-known 22° circular halo often seen around the Moon on a dry winter night. When the orientation is strict, the 22° halo is red to its inside and blue to the outside, but the colour is often destroyed by imperfect alignment of the crystals. There can also be a 46° halo caused by a 90° refraction, but this is less frequently seen. By varying the crystal shapes and the extent to which the many crystals are similarly orientated, a wide range of optical phenomena are produced in the light of both the Sun and the Moon. Halos, mock suns, arcs, coronas and iridescent clouds are commonly associated with diamond dust.

Diamond dust is astonishingly reflective of sunlight. If in the laboratory a layer of water with a thickness of as little as *one-hundredth* of a millimetre were turned into tiny ice crystals (diameter about 1 micrometre), the resulting layer of crystals would be very strongly reflective. Indeed, such a layer present everywhere in the upper atmosphere would reflect almost all sunlight back into space. To a space-borne observer the Earth would then appear brilliantly white and featureless, just as the planet Venus does to an observer on Earth.

In Chapter 7 we saw that the loss to the Earth of solar energy caused by reflection (much of it reflection by ordinary clouds) was closely compensated by the heat-radiation traps in the lower atmosphere. But there is no such compensation for diamond dust. The essential point is that the diamond dust adds nothing to the radiation traps, just because so very little material in the form of tiny ice crystals is so exceedingly reflective of visual light.

We need not search any longer for the condition that would cause an ice age. Turning only one-tenth of 1 per cent of the amount of water normally present in the Earth's atmosphere into fine ice crystals would have a catastrophic effect on the climate. And the resulting disaster would not take long to develop. With most of the 64 per cent of incident solar radiation that now penetrates to the lower atmosphere being reflected back by the high atmosphere into space, the temperature of the land would collapse within weeks and the temperature of the ocean within a few years.

The question we must now ask is why diamond dust is not widespread at present. The answer is that water vapour in the atmosphere – more than ample to produce a devastating quantity of diamond dust – does not have a temperature as low as −40°C. The next question is: what is it that now maintains the water vapour temperature above −40°C? To answer this question, I must now define the 'top' of the water-vapour distribution.

There is, of course, no literal top, in the sense of a complete disappearance of all water, because no matter how high one goes in the atmosphere it is always possible to find a few water molecules. By the 'top' I mean the highest level at which the radiation of heat by water vapour into space has a significant effect on the local temperature of the atmosphere. The ability to emit radiation of a particular wavelength always goes hand in hand with the ability to absorb radiation of the same wavelength. This is true for any atom, molecule or material. As I explained in the previous chapter, water vapour is a strong absorber of radiation with wavelengths longer than 20 micrometres. It is this property which produces the critically important water-vapour trap in the lower atmosphere. In the higher atmosphere, water

127

vapour picks up heat from the surrounding air and proceeds to radiate energy strongly at the same wavelengths – those longer than 20 micrometres. If the water reabsorbs its own radiation (as it has a strong tendency to do in the lower atmosphere), no energy is lost. But high enough in the atmosphere the amount of the water vapour becomes so small that reabsorption begins to fail, so that a considerable fraction of the emitted heat escapes out into space. This is what I mean by the 'top' of the water vapour.

Astronomers define the 'surfaces' of stars in just the same way. The surface of the Sun that we see by eye occurs at the level of the atmosphere where the solar material 'lets go' of the light and heat. To a space-borne observer, operating with wavelengths longer than 20 micrometres instead of with visible light, the surface of the Earth would not be the surface as we understand it at all. The space observer's 'surface' would be the atmospheric level that I am defining as the 'top' of the water-vapour distribution.

As the top of the water-vapour distribution radiates strongly into space at wavelengths longer than 20 micrometres, the energy so lost has to be made good by the surrounding atmosphere. If the surrounding atmosphere had no access to energy, the continuing radiation by the water vapour would cool the air further and further until eventually it fell to 0K, when there would be no more heat to radiate. But then, as the temperature fell through −40°C (233K), diamond dust would form from the water vapour and a climatic disaster would be upon us.

Water vapour is a kind of scientific Jekyll and Hyde. In providing a warming radiation trap in the lower atmosphere, it plays the benign role of Dr Jekyll; but, in its demands for energy in the higher atmosphere, and in its threat to form a reflecting layer of diamond dust, it reverts to the character of Mr Hyde. To avert disaster, it is essential that there is a continuing supply of energy into the higher atmosphere in order to offset the water vapour's continuing radiation of heat into space. The higher atmosphere gains a little energy from absorption of sunlight, especially at heights between 25 kilometres and 50 kilometres above ground, where ozone absorbs ultraviolet light from the Sun. But this is *above* the top of the water-vapour distribution and it is also inadequate. The inescapable conclusion, therefore, is that the atmosphere at the level of the top of the water-vapour distribution must receive a continuing supply of energy from below. That supply is carried by the third 'personality' of water vapour – latent heat.

We can now make the most crucial calculation of all. How much latent heat is needed to prevent the temperature of the radiating water vapour high in the atmosphere from falling below −40°C and therefore prevent the formation of the deadly diamond dust? For the purpose of this calculation, I will suppose the water vapour to be a highly efficient radiator of heat at all wavelengths longer than 20 micrometres, and I will express the required latent heat in terms of the annual amount of evaporation from the ocean that is needed as a vehicle for transporting the latent energy from the lower to the upper atmosphere.

The necessary amount of latent heat turns out to be that given by condensing 56 centimetres (22 inches) of rain each year everywhere on the Earth. (Since, after condensation, the water eventually falls back to the Earth's surface as precipitation, this is an intuitively understandable way to express the result of the calculation.) If the annual precipitation averaged over the Earth were less than 56 centimetres, there would be insufficient upward transport of latent heat to prevent the temperature at the top of the water-vapour distribution from falling below −40°C, diamond dust would form and the Earth's present equable climate would collapse in a few months.

Carbon dioxide also increases the cooling of the upper atmosphere and the effect of additional cooling increases the need for latent heat. If the top of the carbon-dioxide distribution, defined in a similar way to the top of the water-vapour distribution, occurred at just the same atmospheric height as the water vapour, the necessary evaporation rate at the ocean surface would have to be increased by about 50 per cent. The average precipitation would therefore have to be increased from 56 centimetres of rain to about 84 centimetres. The top of the carbon-dioxide distribution lies, however, at a greater altitude than the top of the water-vapour distribution, and this appreciably reduces the importance of carbon dioxide. If we take an estimate of 65 centimetres of rain as the annual average of precipitation required to keep diamond dust at bay, we shall not be far wrong.

Figure 37 shows a world-rainfall map with contours of equal annual precipitation, with the numbers given in centimetres of water equivalent. A word of caution is necessary about these numbers. It is wrong to jump to the conclusion that, wherever a local region of Figure 37 has a rainfall less than 65 centimetres, diamond dust must form there in the high atmosphere – because the winds tend to even up the atmospheric condition, especially those that blow strongly west to east over belts of latitude. For example, high precipitation in Africa and South America easily compensates for the low precipitation in the South Atlantic. The desert regions common at latitudes 30° north and 30° south are protected by a similar effect. Northerly latitudes up to about 70° have precipitation values that are moderately below 65 centimetres – 50 centimetres is typical – but once again the natural heat engine of the Earth, directed from the equator to the polar regions, covers the small difference. Only in the extreme polar caps does it seem that the calculated supply of latent heat is seriously inadequate, and it is of course in the polar caps where ice sheets and glaciers are found, and where diamond dust occurs.

To be able to compensate effectively in the manner described in the preceding paragraph, the average world precipitation must exceed 65 centimetres of rain. Otherwise, the heat engine would spread a deficit everywhere, leading to large areas where diamond dust would form. It is therefore important that the average precipitation throughout the world is actually about 80 centimetres of rain above, but not greatly above, the necessary minimum. We can now jump to an almost obvious conclusion. Ice ages occur whenever the average world precipitation falls below 65 centimetres.

The world precipitation is determined overwhelmingly by the heat content of the ocean, which is why the cooling of the ocean after about 20 million BP was so crucial. The warm ocean of 50 million years ago would have had no difficulty in maintaining an average precipitation far above 65 centimetres. If a powdering of diamond dust arose in the high atmosphere for some reason, the land areas would cool quickly and, with a high-temperature heat source in the ocean and a low-temperature heat sink on the land, a powerful engine would supply so much latent heat into the high

Figure 37: *Distribution of mean annual precipitation at the present day. (After H. H. Lamb, Climate: Present, Past and Future, vol. 1, Methuen, 1972.)*

atmosphere that the ice crystals there would soon be evaporated. When the ocean is warm, therefore, the Earth is well protected against the possibility of an ice epoch.

A cold ocean leads to dismal conditions. The heat engine from the equatorial regions towards the poles has only a feeble drive. Evaporation and precipitation are low. Diamond dust forms in mid-latitudes as well as in the polar regions. Both the ground and the sea in those regions experience a marked depletion of sunlight. Although precipitation on the land is low, it is snow not rain which falls, because of the great cold. The snow becomes gradually compacted into ice; the ice does not melt; over the millennia, glaciers and ice sheets grow slowly to considerable depths. The lack of sunlight at the ocean surface maintains the cold. The cold ocean generates the diamond dust and the diamond dust ensures that the ocean remains cold. The Earth is locked into a self-maintained ice-age cycle from which it is unlikely to be released without some exceptional event.

There is a modest but welcome measure of stability in the Earth's present state. Because the ocean today has about a ten-year store of solar energy, it is inevitable that any short-term atmospheric cause of cooling must affect the land more than the sea – for example, a change in the pattern of anticyclones causing a cooling of mid- and high latitudes. The effect is to maintain the temperature of the heat source in the oceans but to lower the temperature of the heat sink, increasing the efficiency of the engine and so moderating the original cooling. Indeed, as long as the world ocean had enough heat resource, any unusual cooling of any local part of the Earth will be cancelled out by the increased activity of the engine that transfers heat to the deprived regions.

Short-period fluctuations of climate occur constantly from year to year and, in confirmation of the argument of the preceding paragraph, the Earth has avoided returning to an ice-age condition for 10,000 years. The Earth is not, however, protected against a sudden downward step in the heat storage of the ocean. If that happened swiftly, the evaporation of water vapour would be cut, disastrously reducing the supply of latent heat into the high atmosphere. The efficiency of the heat engine would be lowered, as though the pressure of steam in the boiler of an old-style locomotive were lowered, or an inert liquid put together with petrol in the tank of a car.

It is now clear exactly what we are looking for in our search for the cause of ice ages: a process that can rob the ocean suddenly of a significant fraction of its heat storage. If all the 30 million cubic kilometres of ice now in the Antarctic broke into small fragments and melted into the tropical seas, the melting would reduce the heat available in the ocean by about 40 per cent. That is the measure of cooling that we are seeking.

In 1964, A. T. Wilson pointed out that melting at the base of the Antarctic glaciers due to the trapping of geothermal heat (see Chapter 3) is leading to a surge condition there. A surge takes place when a glacier with a small degree of basal slip suddenly loses traction with its bed, so that the bottom part of the glacier is forced outwards. In the late Ordovician, glaciers in north-western Africa appear to have surged in this way from rough land on to the smooth, submerged sandy bottom of a continental shelf. It is not unreasonable to suppose that a similar surge condition could occur around the rim of the modern Antarctic continent.

Dr Wilson believes that this would be enough to set up a new ice age. While I wish him the best of luck with his theory, I do not believe it is the correct one. Clearly, the whole of the Antarctic ice could never surge into the sea in a single year. At the present rate of calving of the Antarctic glaciers and ice sheets into icebergs, it would take 20,000 years for the whole of the ice to pass into the sea. Even if the present rate were to increase quite grossly – say, to a rate at which all the ice disappeared in only 100 years – the annual cooling effect on the ocean would still amount to only 0.4 per cent of the available heat. The Earth's heat energy would only have to work just a little less vigorously, by carrying a little less oceanic heat towards the poles, for the cooling effect to be compensated by a slightly increased rate of storage of solar energy in the tropical oceans.

The melting of icebergs provides water with little or no dissolved salt in it – water that is less dense than seawater. The cool meltwater tends therefore to form the uppermost layer of the ocean, and so it could be argued that, with the uppermost layer of the ocean very cold, the critical evaporation of water vapour might be greatly reduced. But wind action works against this effect, mixing surface water down to a depth of perhaps 100 metres. In this connection, it should be remembered that most of the available oceanic heat is contained in a surface layer only a few hundred metres deep.

Even if one ignores wind action and assumes the fresh melt-water to form a very thin surface layer, the argument is still not adequate. If the Antarctic ice were to melt away in only a century, sea level would rise by rather less than 1 metre per year, as that would be the thickness of the layer of fresh water added annually at the ocean surface. Since sunlight is mostly absorbed in the top few metres of the sea, the freshwater would receive the full benefit of the solar energy falling on the ocean. A straight-forward calculation shows that cold meltwater, initially at 0°C, would heat up in equatorial and mid-latitudes to 20°C in less than a month. I doubt, therefore, that an increase in the calving rate of the Antarctic ice could be a significant factor in the cause of ice ages.

A more likely possibility would be for the ocean to overturn itself, with the cold bottom water coming to the surface. This would certainly have the effect we are looking for – an immediate steep decline in evaporation, with a consequent marked drop in the average world precipitation rate. The formation of diamond dust in the high atmosphere would then plunge the Earth into an ice age. But how could the ocean overturn itself?

A similar process on a local scale produces hurricanes. It is possible to work a heat engine between the comparatively high temperature of the uppermost surface layer of the tropical ocean and the much colder water that lies below; and this is just what a hurricane is – a heat engine. Once it is set up, the engine simply runs itself, provided the hurricane stays over the sea. But, if it should move from the sea to the land, the hurricane loses its energy source. Like a moving car that runs out of petrol, it rolls on for a while and then stops.

All that is lacking from such a theory is a trigger to set a world-wide heat engine going on a vastly greater scale than a hurricane, a heat engine that systematically overturned the whole tropical region of the world ocean. While it is possible to think of various potential triggers, such as a giant meteorite from space crashing into the ocean, or some enormous submarine volcanic outburst, or perhaps even a sudden large surge of Antarctic ice into the ocean, I have not been able to convince myself that any of these eventualities would cause the onset of an ice age. It may be so, but I do not see how to prove that it must be so.

I am inclined, therefore, to pursue the idea of a loss of sunlight caused by a layer of reflective particles. Very little material in the form of fine particles suspended in the high atmosphere would have a devastating effect on our climate. Only one ten-millionth part of the amount of Antarctic ice would be ample to produce such an effect.

The relay, a double switch, is an ingenious device widely used in the electrical industry. The first switch is delicate and can be operated easily and safely by the touch of a finger, or by a small electric current. The closing of the first switch causes a moderate current to flow which acts to close the second switch. With the second closure an exceedingly large current is made to flow – for example, the huge current we use in the mains electricity supply. In this way, a very small initial current – or the touch of a finger – can be used to control the disposition of a very large amount of energy. Fine particles in the upper atmosphere work in a similar way. They can control the behaviour of many millions of times more material at the surface of the Earth.

Fine particles have another interesting and significant property. They do not stay up aloft indefinitely. Gravity pulls them downwards through the air. Large, heavy particles fall quickly, while very small ones remain suspended for much longer. It is remarkable that particles of just the sizes most effective for the reflection of sunlight – that is, about half a micrometre – take about ten years to fall through the high region of the atmosphere (altitudes above about 12 kilometres). Small particles fall more quickly below 12 kilometres, in spite of the increasing density of the lower air, because they become incorporated into much larger ice crystals and water droplets whose greater weight causes them to fall faster through the air.

As we have seen, the ocean has just about a ten-year supply of solar energy. Any process that blocked or seriously reduced the intensity of the sunlight reaching the ocean surface for an interval of ten years would, therefore, have just the effect we are seeking.

135

It would run down the heat storage in the ocean to the point where diamond dust forms in the atmosphere. The diamond dust would then complete the plunge into an ice age.

The mystery is now reduced to a search for some process which injects a sufficient quantity of small particles into the region of the atmosphere at altitudes above 12 kilometres, the region known as the stratosphere. There are three processes to be considered: (1) the explosion of volcanoes; (2) micrometeorites entering the Earth's atmosphere from space; (3) the violent collision of great meteorites with the Earth.

An article by Henry Strommel and Elizabeth Strommel in the issue of *Scientific American* for June 1979 explains the effect of volcanic explosions:

In New England, Canada and Western Europe the summer of 1816 was extraordinarily cold. A meteorological record for New Haven that had been kept by the presidents of Yale College since 1779 records June, 1816, as the coldest June in that city, with a mean temperature that would ordinarily be expected for a point some 200 miles north of the city of Quebec. The Lancashire plain in England had its coldest July, and the summer as a whole ranks as the coldest on record in the Swiss city of Geneva for the entire period from 1753 to 1960. In New England the loss of most of the staple crop of Indian corn and the great reduction of the hay crop caused so much hardship on isolated subsistence farms that the year became enshrined in folklore as "Eighteen Hundred and Froze to Death". The calamity of 1816 is an interesting case history of the far-reaching and subtle effects a catastrophe can have on human affairs.

The chain of events began in 1815 with an immense volcanic eruption in the Dutch East Indies (now Indonesia) when Mount Tambora on the island of Sumbawa threw an immense amount of fine dust into the atmosphere. Sir Thomas Stamford Raffles, who commanded a British military force on the island, described the eruption in his *History of Java*: 'Almost every one is acquainted with the intermitting convulsions of Etna and Vesuvius, as they appear in the description of the poet and the authentic accounts of the naturalist, but the most extraordinary of them can bear no comparison, in point of duration and force, with that of Tomboro. This eruption extended perceptible evidences of its existence . . . to a circumference of a thousand statute miles from its centre, by tremendous motions and the report of explosions; while within the range of its more immediate activities, embracing a space of three hundred miles around it, it produced the most astonishing effects, and excited the most alarming apprehensions. On Java, at a distance of three hundred miles, it seemed to be awfully present. The sky was overcast at noonday with clouds of ashes; the sun was enveloped in an atmosphere, whose 'palpable' density he was unable to penetrate; showers of ashes covered the houses, the streets and the fields to a depth of several inches; and amid the darkness explosions were heard at intervals, like the report of artillery or the noise of distant thunder. So fully did the resemblance of the noises to the report of cannon impress the minds of some officers, that from apprehension of pirates on the coast vessels were dispatched to afford relief.'

This eruption, which was considerably larger than the better known one of Krakatoa in 1883, reduced the height of Mount Tambora by some 4200 feet and ejected some 25 cubic miles of debris. Ash was encountered by ships at sea as large islands of floating pumice as much as four years after the event. Climatologists rank the eruption as the greatest producer of atmospheric dust between 1600 and the present. The dust circled the Earth in the atmosphere for several years, reflecting the sunlight back into space and thereby reducing the amount of it reaching the ground.

The idea that dust in the upper air can result in lower temperatures at ground level is quite old. Benjamin Franklin invoked it to explain the cold winter of 1783–84. Today the idea can be confirmed more conclusively through long records of temperature from many parts of the world, which can be compared with the fairly complete record of the volcanic eruptions that have been observed during the past two centuries.

Here is a clear-cut experimental proof of the existence of the effect we are seeking – Mount Tambora did the experiment for us

137

– but a notable volcano erupts every few years and not one of them has brought on a new ice age. It is possible that past ice ages were caused by volcanoes bigger than any which has erupted over the past ten millennia. There is no direct disproof of this postulate but, since we have no direct evidence to support it, the other two of our three possibilities must also be considered.

Micrometeorites produced in the break-up of comets enter the Earth's atmosphere all the time. A considerable fraction of these small particles are the right size to reflect sunlight efficiently; each year a total quantity of perhaps 10,000 tons of micrometeorites are added to the atmosphere. However, this process, just like the eruption of volcanoes, is incapable now – and it must have been incapable over the past 10,000 years – of producing an ice age. The addition rate needed to produce an ice age would be about 100 million tons per year – very much greater than the present rate. It would therefore be necessary to postulate that there can be exceptional occasions when the addition rate rises very greatly above the present rate, for which, like the volcanic hypothesis, we have no direct evidence.

So let us come now to the third of the above processes: the collision of a giant meteorite with the Earth. This happens. At the landing point of the meteorite, the collision produces a crater similar to those found in great profusion on the surface of the Moon. Figure 38 shows the positions of the larger meteorite craters that have been found on the Archean shield of Canada. Smaller craters are more widespread than those of Figure 38. R. Grieve of the Canadian Department of Energy, Mines and Resources, has estimated that about 5000 giant meteorites with diameters of more than a kilometre have hit the Earth over the past 600 million years, with an average strike rate of one per 120,000 years. Meteorites with diameters greater than 300 metres must hit the Earth once in every 10,000 years.

Let me explain why the size of the meteorite is important. A tiny meteorite, a micrometeorite, experiences a sudden flash heating as it enters the Earth's atmosphere. For a typical particle, 1 micrometre in size, the temperature rises to about 500K (227°C) over a period of one or two seconds. The flash heating becomes stronger as the size increases, until for a meteor the size of a pinhead the temperature becomes sufficient to vaporize it completely. It becomes a trail of hot gas through the atmosphere. The hot gas emits light and, if the incident occurs at night, an observer on the ground sees the streak of light and calls it a 'shooting star'. If the meteor is pea-size rather than pinhead-size, the amount of evaporated gas is more. The streak of light is far more brilliant.

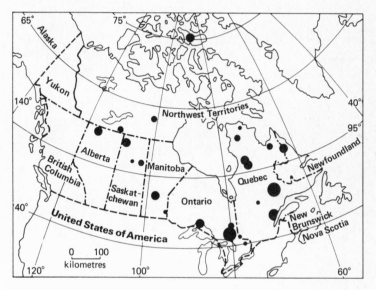

Figure 38: *Distribution of craters thought to be of meteoric origin in Canada. (After R. Grieve, Canadian Department of Energy, Mines and Resources.)*

If the incoming body is the size of a clenched fist, the air pressure on the body is large enough to slow down its original very high speed, and this happens before the heat can evaporate the whole of the material. A residue remains which cools and falls to the ground at a comparatively low speed. The body is now called a meteor*ite*, which means that a part of it has managed to reach the ground. The pressure exerted by the air on an incoming meteorite grows with the size of the meteorite. When the body is a few metres in size, the pressure becomes large enough to burst

139

it into fragments, which then fall to the ground as a meteorite shower.

With another considerable increase of scale – to a size of several hundred metres – the body no longer bursts into fragments which can be treated separately, because it is now so large that there is insufficient time for the fragments to fly apart from each other. The body tends, therefore, to act as a coherent whole, and the air pressure – although great – is insufficient to slow the very high incoming speed. The body crashes down through the atmosphere to make an explosion pit or crater at the place where it lands. The explosion causes material torn from the ground to be splashed outward and upward with the disintegrating meteorite, radiating from the point of impact.

The most violent meteoritic event of modern times occurred just after midnight on 1 July 1908. Miss K. Stephen of Godmanchester, Huntingdon, wrote to *The Times* about a strange light she had seen in the sky, commenting that 'it would be interesting if anyone could explain the cause of so unusual a sight'. The explanation was a long time in coming. Indeed, it was a long time before the point of impact was found; it was not until 1927 that an expedition under L. A. Kulik penetrated to the region of the Tunguska river in Siberia, to discover a scene of peculiar devastation. An explosion crater might have been expected, but there was none. Instead, there was an extensive area of flattened trees with a small plantation of branchless, standing trunks at its centre. No meteorite fragments were found. There have been wild and fanciful suggestions to explain these somewhat bizarre facts. The most likely explanation is that the meteorite, while not large enough to crash through the atmosphere, had an exceptionally high speed of entry from space. Entry speeds of 30 kilometres per second are typical, but in exceptional cases the speed can be as high as 70 kilometres per second. Each ton of material then has some five times more energy than the typical bodies. It seems that, so great was the heat released by the Tunguska event, that the whole meteorite simply evaporated after bursting into fragments. The explosion occurred in the air itself, not at the surface of the ground, and it was the resulting enormous air blast, comparable to an air blast from a very large nuclear bomb, which devastated the surrounding forest for scores of kilometres around the place of entry.

A giant meteorite of the much larger size I have suggested – 300 metres or more – would certainly crash to the ground and would produce very much greater devastation than the Siberian meteorite of 1908. Even with a typical entry speed of 30 kilometres per second, the energy per ton of meteorite is sufficient to throw 1000 tons of debris up to a height of nearly 50 kilometres in the atmosphere, which is to say to the top of the stratosphere. A spherical meteorite 300 metres in diameter weighs about 50 million tons and is capable of throwing 50,000 million tons of debris up into the stratosphere. Only a few per cent of the debris, in the form of tiny dust particles, would be ample to cause an ice age. The dust particles would spread gradually over the Earth, producing a blanket of reflective particles throughout the high atmosphere. The amount of sunlight reaching ground level would then be reduced appreciably until the small particles fell down through the stratosphere – taking a period of about ten years. This could well be the relay switch we have sought.

A giant meteorite is potentially more dangerous than a volcano because, although the amount of material in a giant meteorite is no greater than the amount blown off by a volcano, each ton of meteorite has vastly more energy than each ton of material from a volcano – about a thousand times more. A giant meteorite is, therefore, capable of spraying up into the stratosphere very much more debris than a volcano can, at any rate the volcanoes of which we have experience.

The nature and origin of giant meteorites are interesting. The following passage is abstracted from an article of G. W. Wetherill which appeared in the March 1979 issue of *Scientific American*:

In 1937 a body about a kilometre in diameter, later named Hermes, passed within 800,000 kilometres of the Earth, no more than twice the distance of the Moon. It has not been seen again. About once in every century a similar object can be expected to travel past the Earth at less than the lunar distance. And once in every 250,000 years, on the average, the

141

Earth and such a body will collide. The impact of the collision will release energy equivalent to 10,000 ten-megaton hydrogen bombs and will make a crater some 20 kilometres in diameter.

Hermes, the asteroid discovered crossing the Earth's orbit in 1937, was only the third object of its class to be identified. The first such object had been discovered earlier by Karl Reinmuth of the University of Heidelberg in the course of a photographic search for ordinary asteroids.

Ordinary asteroids vary in size. The smallest is only a few metres in diameter, while the largest one, named Ceres, has a diameter that is about a quarter of that of the Moon. Ordinary asteroids move round the Sun in more or less circular orbits, as the Earth does, but their orbits are generally about two and a half times larger than that of the Earth.

The object which Karl Reinmuth discovered, and which he named Apollo, was an asteroid with a quite different kind of orbit. It was far more elliptical than the orbit of an ordinary asteroid or that of a planet. The essential characteristic of the class of asteroid named Apollo-objects is that, over a part of their orbits, they come closer to the Sun than the Earth ever does and, over other parts, they move farther out from the Sun than the Earth ever does. This makes it possible for them to collide with the Earth, which they do from time to time on a random basis – about once in 250,000 years, according to the astronomical estimates; and about once in 120,000 years, according to an actual count of craters on the Earth's surface.

The only other objects in the solar system with similarly flattened orbits are the comets. Largely because of this similarity, it is believed that the Apollo-objects have evolved from comets. They are residues from comets, what is left when all volatile substances (such as water) originally present in the parent comets have been evaporated by the heat of the Sun.

Gradually, over the years, more and more Apollo-objects have been discovered, and in 1973 the first systematic programme for their detection was started at the California Institute of Technology by E. M. Shoemaker and E. F. Helin. Other programmes have followed, set up by C. Kowal in California and by astronomers of the European Southern Observatory in Chile. From his survey, Shoemaker has estimated that, above a certain threshold of photographic brightness (corresponding to a diameter of about 1 kilometre), there are about 750 Apollo-objects. The number of smaller objects, say with a diameter of 300 metres, is probably some ten times greater, say about 10,000. These are the objects capable of initiating a new ice age.

To quote Dr Wetherill again:

> Until recently scant attention was paid to [Apollo] objects even by the minority of astronomers specializing in planetary studies. It was only gradually recognised that these small asteroids have an importance to the Earth and planetary science quite out of proportion to their size and number. Bodies of the Apollo-type have been the principal producers of craters larger than 5 kilometres in diameter on the Earth, Moon, Venus and Mars (with a possible reservation in the case of Mars).

The reason for making Mars a possible exception is that ordinary asteroids do occasionally collide with it, so that ordinary asteroids may have wreaked more havoc on the surface of Mars than the Apollo-objects have done. Ordinary asteroids must collide fairly frequently with Apollo-objects. Bits then get knocked off both kinds of asteroid. Bits from the Apollo-objects usually continue to move in highly flattened orbits around the Sun, following much the same form of path as their parents. The bits are therefore also subject to collision with the Earth and, because they have become far more numerous than the Apollo-objects themselves, minor collisions happen far more frequently than the full-scale crashes of Apollo-objects and the Earth. Such bits of asteroids are the smaller meteorites that enter the Earth's atmosphere every few months or years.

Wind and water erosion on the Earth soon remove evidence of most of the smaller craters produced by collisions of Apollo-objects with diameters at the lower end of the range – say 300 metres or less. Such small craters may be preserved in exceptional circumstances, however, particularly in desert regions such as Winslow, Arizona, where the well-known meteorite crater is estimated to be between 25,000 and 50,000 years old. It

143

is thought to have been formed from an iron object about 100 metres in diameter.

Because of weathering, only the largest craters survive at the Earth's surface for longer than 500,000 years. The Clearwater Lakes of northern Quebec lie in a crater 30 kilometres across – that is an example of a very large crater. Scientists have estimated that the whole land area of the Earth has been subject to about 130,000 major impacts over the past 1000 million years. That gives an average strike rate of one impact in about 10,000 years – remarkably similar to the timespan involved in the ice ages. The graphs drawn from the pollen and ocean-core data (Technical Note 3) show that sudden shifts of climate occurred at discrete moments; the intervals are irregular, but the average spacing of the zigs and zags is about 10,000 years.

There can be no doubt about the reality of the impact of giant meteorites. If it were not for rapid weathering, the surface of the Earth would be just as crater-strewn as the surface of the Moon. Such an impact is indeed the event that could jerk the Earth from its present warm cycle into an ice-age cycle. But for the ice-cold water from the Antarctic, the heat storage in the ocean would last for fifty years – considerably longer than the reflective particles would take to fall from the stratosphere to the ground. As it is, only the heat capacity of the top 500 metres of ocean water is available to feed the land, and the engine would run out of heat in ten years or less, *before* the reflective particles had fallen from the stratosphere. The ocean could no longer supply latent energy to the atmosphere. Diamond dust would form and remain to cut off the sunlight after the particles had fallen to the Earth's surface. The diamond dust would keep the ocean cold and the cold ocean would have no heat to remove the diamond dust. A new ice age would have arrived.

9 The Beginning of an Ice Age

The really exceptional aspect of this theory of the cause of ice ages lies in its prediction of a ten-year precursor period – a period of bitter cold on the land and tremendous storms from the sea. Deprived of sunlight while the particles are suspended in the atmosphere, the land cools much more quickly than the ocean. In other words, the loss of energy into the sink of the Earth's heat engine is greatly increased. Consequently, the engine works more powerfully for a while, draining the ocean of its heat at an increased rate. As long as the heat storage in the ocean lasts, storm activity becomes more and more widespread and violent.

I suspect that it must have been during the catastrophic ten-year period at the onset of the last ice age that many of the erratic boulders discussed in Chapter 3 were split from their rocky outcrops and distributed widely about the landscape. The traditionally accepted theory is that erratic boulders were separated from their parent rocks by frost action. Water trickles over rocks, penetrating into joints and cracks where it freezes each winter, or several times each winter. As it freezes, the water expands and, in doing so, exerts a large force which tends to split the rock, enlarging the cracks and allowing more water to accumulate within them. Freezing and refreezing widens the cracks until eventually a boulder breaks off and rolls down into the valley below under gravity. This process requires a climate with alternating cold and warm spells because the more often the water freezes, thaws and refreezes, the more effective the frost action would be. The winter climate of the British Isles today, with its almost daily alternation between warm and cold, is ideal for this kind of frost action, but it produces only minimal quantities of boulders compared to the amount generated during the last ice age.

I became suspicious of the claims for frost action in 1966 after the great storm which hit the Lake District that August. I did not witness the storm myself, but a friend told me that, after darkness had set in, there had hardly been a moment when the eye was not aware of a lightning flash. The havoc caused by the storm was the talk of the district. For example, between Borrowdale and Langdale, a large quantity of new boulders, each about one foot in size, had been split from Sergeant's Crag. Obviously, frost action could not have been responsible for this in mid-August, and in any case all the boulders had been produced in a single day. The local suggestion was that they had been caused by water action, and indeed much of the gravel in the area must have been washed there by grossly swollen streams. But there is notable absence of water on the slope of Sergeant's Crag. To have suggested that the large number of new boulders at its foot had been scoured and polished in a stream bed was obviously ridiculous. There could be no explanation for them other than lightning – lightning flashes which had been so closely spaced as to seem continuous.

In its entry on 'Lightning' the *Encyclopaedia Britannica* states that:

> Large pressures are generated when lightning strikes objects, instantly evaporating water in cracks to split trees so as to cause explosions. At 100,000 ampères, blocks of stone weighing five tons may be torn loose and rocks weighing 50 pounds may be thrown 20 yards or more.

A spherical rock weighing 50 pounds has a diameter of about 10 inches (25 centimetres), and a large number of such rocks generated at the top of a steepish slope and exploded with speeds that would propel them for 20 yards (20 metres) on flat ground would indeed lead to their being scattered down the slope.

The feature of Sergeant's Crag that distinguishes it from other hills in the area (which were not strewn with boulders) is that it is a little knob of a thing, an eminence, the natural target for lightning. Since 1966, I have verified that lightning is a likely source of much talus on British hills. I have always found a rocky eminence above slopes rich in small boulders. The essential property of the eminence is not that it is particularly high but that it is the natural strike point for lightning. It can be as little as 100 feet higher (30 metres) than the surrounding countryside, provided there are no other eminences nearby to compete in attracting the lightning. In this respect, eminences of middle height like Sergeant's Crag often command a wider area than the more obvious projections near mountain tops. On higher ground, there is usually a multiplicity of targets for lightning, so that strikes near mountain tops tend to be widely shared between competing projections.

During the ten-year onset of the last ice age, storms similar to that of August 1966 would have been commonplace, almost daily, occurrences. As the heat engine supplied energy to the cooling land through the latent heat of water vapour, the release of the water vapour would have caused violent storms and almost continous lightning. The quantity of detritus produced could have been vast, and it could have come largely from the eminences which attracted most lightning. Among eminences without competitors in their own areas were Ailsa Craig (Figure 13), the source of the Shap granite erratics and scores of other seemingly trivial small rises like those shown in Figure 39.

The storms must also have involved enormous quantities of precipitation. We have seen that it is the evaporation of water vapour from the ocean that drives the heat engine. With the engine working at exceptional intensity during the catastrophic period, much of the ocean's heat would have gone into evaporation. The heat available in the ocean today is sufficient to evaporate a surface layer about 30 metres deep. With much of the latent heat in the water vapour being directed towards the land in mid- and high latitudes at the beginning of the ice age, the precipitation rate there would have been very large indeed, just because the area of the ocean is so much larger than the area of the land. A precipitation rate on the land of 10 metres per year throughout the catastrophic period is therefore by no means an excessive estimate.

The latent heat released by the condensation of so great a quantity of water would ensure that much of the precipitation would occur as rain. Freshwater lakes would fill to great heights, particularly as freezing on the ground would block their outlets with massive and ever-growing accumulations of ice. Water falling on very cold ground would soon freeze into smooth ice,

147

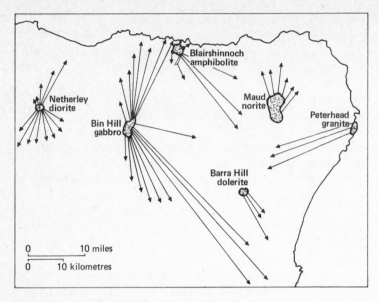

Figure 39: *Distribution of erratics in Aberdeenshire.*

covering minor irregularities, stones and boulders. Before the end of the catastrophic period, however, the layer of ice – always smooth at its surface – would rise higher and higher to become a sheet several hundreds of metres thick. The ice would become an enormous skating rink stretching for thousands of kilometres over the land, perhaps the whole way from north-western Europe to distant Siberia. Precipitation would be still greater on the mountains, where glaciers would accumulate at an alarming rate of perhaps 100 feet (30 metres) a year. The ice of the mountain glaciers would be in a violent, almost turbulent condition, capable, I would suppose, of exerting a most drastic grinding effect on the whole mountain landscape.

This rapid accumulation of ice into a huge skating rink would tend to cover small eminences, but lightning can tear ice even more readily than it tears rock. While ground which attracted no lighting – or only an occasional flash – would become ice-covered, raised eminences which were struck repeatedly would be

148

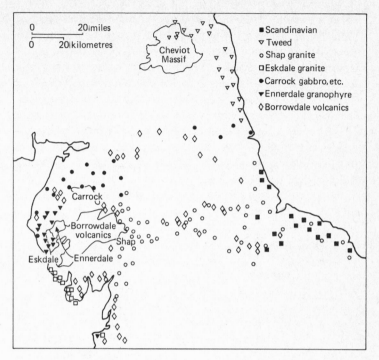

Figure 40: *Distribution of erratics from Scandinavia and from the Lake District in northern England.*

stripped of their ice. Boulders and smaller debris would then be struck from them to fall on to the surrounding smooth ice and slide away downhill. Thus, debris at Shap would tend to slide away to the north down into the Eden Valley, and to the south down on to lower ground near Kendal. A picture of the diffusion of erratic boulders is beginning to emerge, but we do not yet have an explanation of the very wide distribution of boulders, such as the spread of Shap erratics to the Yorkshire coast (Figure 40).

Many times in the hills one encounters a stiff wind. It is easy on such occasions to discover that, by facing into the wind and leaning on it, one can maintain a stable position with the body inclined at an angle of 30° or more to the vertical. It is clear that a stiff wind produces a horizontal force on the body about a half as

149

strong as gravity. Horizontal wind forces of a half, or a fifth, or even a tenth, of gravity would be sufficient to move boulders on smooth ice. Even less force would be needed if the boulders themselves became encrusted in ice, since there is very little friction indeed between one icy surface and another. A boulder with a horizontal force of one-third of gravity on it could be pushed uphill on slopes with gradients up to about 18°. The route taken by Shap granite, along the line of the Roman road from Brough to Barnard Castle, involves nothing like such a slope. The line of the Roman road might just as well have been flat for wind-impelled boulders. There is no difficulty at all in understanding how boulders could slide great distances, even uphill.

My estimates of forces and angles apply to boulders a foot or two in diameter (50 centimetres or so) – about the same size as a crouching human. The horizontal forces on much larger boulders would of course be appreciably less, decreasing as the diameter increased. A boulder 20 feet (6 metres) in diameter would be significantly more limited in the slopes it could ascend; even if it were encrusted in ice, it could only ascend a slope of 1° or 2°.

Boulders would have to travel quickly. If they lay around for weeks on end waiting for a suitable wind, the heavy precipitation and the consequent rapid accumulation of ice would soon cover them over. They would be whisked away as they were split off, to end up wherever the vagaries of the wind deposited them, then to become entombed in ice. Since winds would blow from different directions on different days, the boulders would sometimes slide north, sometimes south, or east or west. This is the reason for the generally radial pattern of boulders from every minor eminence shown in Figures 39 and 13 (page 50). We are no longer required to accept the really quite absurd notion that a glacier spread in all directions from Ailsa Craig; it was simply that, over a period of several years, there were days on which the wind blew across Ailsa Craig in every direction. Nor do we need to worry that the tracks of boulders from one eminence crossed over the tracks of boulders from other eminences; the situation was no different from that of an ice rink with skaters approaching each other from all directions. Mention of ice rinks reminds me of a curious coincidence: Ailsa Craig granites are used for curling stones.

Let me confess that I have glossed over quite a problem, by assuming in the above discussion that water falling on to cold land produced smooth ice at all latitudes. Ice ages are not all-or-nothing affairs that glaciate the entire Earth. The supply of latent heat to the atmosphere is so large in low latitudes that, even with a run-down ocean, the tropical zone would probably remain substantially free from diamond dust. It is likely that a polar condition would spread towards the equator but would not reach it. The equatorial ocean would continue to be heated by sunlight and the heat engine which it maintained would work to push the glacial condition back towards the poles. If precipitation were heavy enough, the local release of latent heat might be so large that no freezing would occur. Places towards the equator, being nearer the comparatively warm equatorial ocean, would be more likely to have heavy precipitation than regions in higher latitudes. The full story of an ice age lies almost surely in a complex climatic struggle between a still-sunlit tropical ocean and cold, glaciated mid- and high-latitude regions, wan and faded from lack of sunlight.

The British Isles has an equable, marine-dominated climate. This made for a complex situation during the last ice age, when the east–central region of England and the south avoided an accumulation of ice. Figure 2 (page 20) shows the line of demarcation between ice-covered (hatched) and ice-free (clear) regions. The line was not determined by any particularly overriding considerations; it just happened that way in the last ice age, and it happened differently in earlier ones. Irish Sea ice, North Sea ice, and the ice sheet which covered the areas from Cheshire south to Wolverhampton, did not spread out from anywhere. It formed *in situ* as the precipitation became frozen. This locally formed ice was not in any way due to ice spreading from Wales or from the Lake District or from Ailsa Craig. Indeed, Ailsa Craig only comes into the story because smooth ice froze around it, because it was a high point from which lightning dislodged debris, and because extreme gale-force winds blew the debris along the ice to all points of the compass.

Boulders from Scandinavia tell a similar story, but one on a bigger scale. Their tracks cross over in the same way, easy to understand with the curling-stone picture in mind, but impossible

151

with the glacier-movement theory. The low-lying Åland Islands played a role similar to Ailsa Craig's. The islands were exceedingly prolific sources of erratic boulders, which have been found along radial lines to great distances – some indeed as far away as western Russia. There was nothing to stop the Scandinavian boulders being blown farther and farther from their sources except the limit of the ice sheet itself. Since the ice sheet extended further to the west, south and east of Scandinavia than did the ice sheet in the British Isles, the distances achieved by far-journeying Scandinavian erratics were much greater than those of the indigenous British erratics shown in Figure 40.

Drumlins! Whenever I approach Kendal from the south, my heart is lifted by the sight of the drumlins. They are the verdant little hills about a third of a mile long and a hundred feet high which can be seen dotted over the countryside. Any sharp projection in the land would at first stick up through the growing shallow ice sheet of the violent precursor period of an ice age. Such a projection would tend to intercept debris blown by the gales. Under dry conditions, the intercepted material, especially bits of grit and the smaller stones, would eventually swirl away and continue its motion in the wind. But, with heavy precipitation steadily turning into ice, some of the debris would inevitably become firmly bonded to the projection which would tend to grow. There are then two possibilities: the hump of material might grow more slowly than the ice sheet, in which case it would soon be covered by ice and nothing further would happen to it; or the hump might grow more rapidly than the ice sheet. In the second case, the hump would grow into a considerable mound. The issue turns on the local rate of supply of debris. If the supply were fast enough to build one such mound, it would be fast enough to build several, just as it did in the drumlin swarm south of Kendal. The mounds tend to grow longer in the direction that faces into the prevailing wind. Thus the swarm south of Kendal – its shapes are shown in Figure 41 – indicates a prevailing wind either from the north-east or the south-west. If one supposes the main source of debris to be in the north, then the main supply was from the direction of the Howgill fells near Sedbergh.

All these dramatic events occurred in the ten-year catastrophic period at the end of the last ice age, not during the ice age proper, otherwise the sources of erratic boulders – especially in Scandinavia – would have been covered by the great depths of ice which accumulated slowly as the succeeding ice age developed. If the far-travelling erratics with sources near sea level (e.g. the larvikites) had been produced after the ice sheet was formed, they would have had to be carried at the base of huge glaciers, glaciers that could never have moved such great distances. Moreover, the boulders at the glacier bottoms would then have been frequently scored and polished by grinding at high pressure, and boulders like that of Figure 12 (page 49) have not been.

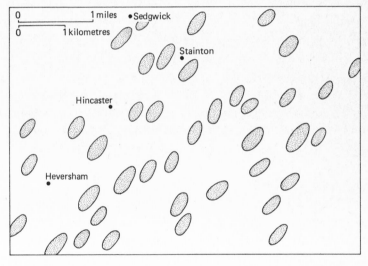

Figure 41: *The drumlin swarm south of Kendal. (After R. Millward and A. Robinson,* The Lake District, *Methuen, 1970.)*

Let me end this chapter with a few reflections more of a cultural than a scientific nature. Our ancestors lived through many appalling climatic episodes – if, indeed, 'lived through' is the correct description. Surely animal life would be totally extinguished in such conditions? Very many individual animals undoubtedly perished but, except in certain cases, some of each species escaped. Killing every member of a species is nearly always an unexpectedly difficult matter, as farmers who seek to exterminate insect pests soon discover.

Human survivors from the traumatic conditions of the onset of an ice age would undoubtedly have attempted to leave a memory of their experiences. Unable to leave a written record, they would have had no recourse except to tradition. Illiterate peoples always guard their oral traditions with the greatest care. Indeed, they must do so if their culture is to have continuity. It is a modern conceit to imagine that nothing meticulously accurate or of high quality can be passed by an oral tradition from generation to generation. Homer's *Iliad* was written in almost the first moment that alphabetic writing became available. The *Iliad* is evidently the product of an earlier oral tradition, and it remains, nearly three thousand years after it was written, the finest story I have ever read.

Even so, there is one important respect in which a documentary account of an exceedingly unusual event is superior to an account handed down by word of mouth. As the generations roll by, people inevitably lose contact with the details of an event that happened long before. Those who continue to tell the story come to understand it imperfectly, with the result that it gradually becomes garbled – garbled first from tradition to legend, and then from legend to myth. A document, on the other hand, may become obscure to a particular generation, but it remains exactly the way it was written, always available for accurate interpretation by a still later generation. The gradual garbling of history is inevitable in an oral tradition, although such is the standard of verbal accuracy of illiterate peoples that garbling is usually a slow process. One can be reasonably certain, therefore, that myths already muddled at the beginning of written history must be truly ancient.

That brings me to the myth of Noah and his Ark. The myth of Noah has commonly been attributed to a purely local flood disaster that overtook early settlers in the Mesopotamian river valleys, perhaps around 5000 BC, and this indeed appealed to me as a sensible explanation when three decades ago I read the well-known book *What Happened in History* by Gordon Childe. Looking at the story now with a more critical eye, I cannot see anything to recommend that interpretation. Around 5000 BC, much was happening to human society – in the Balkans, in Turkey, over a wide sweep of territory from the Nile through Mesopotamia to India and China. I find it hard to credit that what happened to just one group of river-valley pioneers could have given rise to a myth which later spread so widely. With death and privation the common lot of man everywhere, who, except the pioneers themselves, would be interested? Floods are endemic in river valleys, and I cannot believe that among a river-valley people Noah was the only person with a boat.

In the last 10,000 to 20,000 years, the intense and deadly conditions typical of the beginning of an ice age must have occurred at least partially, if not in full measure. Humans over a wide area could indeed have been entirely wiped out, except perhaps for those who survived because of some exceptional advantage. I do not think it can have been a boat – the Ark would be a later garbling of the story, since to later generations a boat would have seemed the logical way to escape from intense rain-fall – but it could well have been a strategically placed cave. I also find it much easier to believe that humans coexisted with other animals in a cave than in a boat.

Both the biblical and the Babylonian versions of the Noah story describe how, when the waters which covered the world at last subsided, the Ark grounded itself on the summit of Mount Ararat. Since the height of Mount Ararat exceeds 16,000 feet (about 5000 metres), one can question the accuracy of this part of the story. The Mount Ararat invention (as it surely must be) is commonly attributed to an attempt by the teller of the myth to give emphasis to the vast extent of the world flood. But this is to confuse the recounting of a myth with storytelling. Myths are transmitted, not invented. I believe that myths change only as those who retell them feel the old story to be in serious error in some way. Changes are then made, not out of poetic licence, but to make the obscure seem understandable. But what could be understandable about the summit of Mount Ararat?

My memory goes back to a day in November 1963 when I stepped out of an hotel in Erevan, Armenia. I looked across at Mount Ararat on the other side of the Turkish frontier. My only thought then was of how hard it would be to climb to the summit.

155

Noting that the mountain was conveniently dome-shaped, I thought it might not be too difficult with crampons. My eye traced a possible route up the shining snow and ice fields. Ice! This was the point where the Ark was said to have grounded when the waters went down. How else could people living in the warm lands of the Near East explain that it was ice that Noah found everywhere around him as he emerged from his cave?

10 The End of an Ice Age

The first short phase of intensive activity is to be contrasted with the marked inactivity of the long later period of an ice age. Once the ocean had been completely run-down, precipitation fell away to low levels. Glaciers continued to accumulate, but only at a rate comparable to the precipitation in Antarctica at present – about 10 centimetres thickness of new ice each year. Glacier motions became ponderous, especially as great cold lowered their plastic flow rates. (In this connection, it should be noted that, while Glen's basic law of plastic deformation may be considered to hold irrespective of temperature, the actual rate of the deformation is strongly temperature-dependent. The rate described in Chapter 3 was consistent with present-day temperatures. At lower temperatures during an ice age the rates could well have been some ten to a hundred times slower than they are now.)

The implication of the picture is that nothing very dynamic happens during the long duration of an ice age. Dynamic action is reserved for its beginning, and indeed for its end. I have explained that the ice-age cycle is self-maintaining, the cold ocean being unable to remove the diamond dust that prevents sunlight from warming the ocean. What, we may ask, can then jerk the Earth back to a warm cycle? The answer is another injection of fine particles into the high atmosphere, but particles of metal rather than of rock. Grains of rock reflect sunlight, whereas particles of iron absorb sunlight more than they reflect it. A layer of iron particles in the stratosphere capable of absorbing only 20 to 30 per cent of sunlight would make the upper atmosphere very warm indeed during the daytime. The temperature of the air would rise far above −40°C, and in a flash the diamond dust would be gone. The remaining 70 or 80 per cent of sunlight would blaze down on the ocean for the time which the particles took to fall from the stratosphere. This would be at least ten years, probably considerably longer. There is an interesting difference between absorptive metallic particles and reflective rock particles. The latter become inefficient reflectors when they are appreciably smaller than 1 micrometre, whereas small iron particles are equally absorptive whatever their size. Since very tiny 'submicron' particles fall from the stratosphere more slowly than micron-size particles do, they can remain suspended, efficiently absorbing heat for several decades.

The ocean would then build a new store of heat, say a ten-year supply, before the particles fell to the ground. The Earth's heat engine would have been refuelled and would return to its work of supplying latent heat to the atmosphere. Further diamond dust would be prevented and the Earth would remain in a warm cycle until the next catastrophic event supplied a layer of reflective particles to the stratosphere along the lines discussed in the previous chapter.

It is satisfactory that both of the switches, to and from an ice-age condition, can arise from the same kind of cosmogonic event. There is nothing in the least contrived about this explanation of the difference between plunging into an ice age and coming out of it, since it is a fact that both iron and stone meteorites do exist. The evidence shows that warm periods have tended to be considerably shorter than glacial periods – generally about 10,000 years compared to the 100,000 year glaciations. In this respect too, the theory is supported by evidence: large stone meteorites are about five to ten times more common than iron ones. It used to be thought that 'irons' were more frequent than 'stones', but this early impression has been found to be false. The casual observer who comes by chance on a fallen meteorite is much more likely to notice a lump of iron than a mere stone. Iron also stands up to erosive influences far better than most of the 'stones', some of which are so friable that they disintegrate in only a few years.

The broad physical picture is now complete. No property of matter has been used that is not well supported by information from the laboratory, and no astronomical event has been assumed that is not supported by clear-cut evidence.[1]

[1] For a more detailed explanation see F. Hoyle and N. C. Wickramasinghe, 1978, *Astrophysics and Space Science*, 53, 523; E. J. Butler and F. Hoyle, 1979, *Astrophysics and Space Science*, 60, 505; E. J Butler and F. Hoyle, 1980, *Moon and Planets*.

Any attempt to analyse an event we have not actually witnessed must suppress details. Overwhelmingly, we tend to concern ourselves with averages. I have said that a blanket of small iron particles in the high atmosphere sufficiently thick to absorb 20 to 30 per cent of sunlight *on the average* would bring an ice age to its end. Locally, however, there would inevitably be places where the absorptive blanket was considerably thicker than its average. Most places in high latitudes would probably experience such a condition temporarily, just as from time to time most places experience abnormally heavy cloud.

This could well explain something that has baffled scientists for generations: the extinction of the woolly mammoth. The mammoth became extinct not at the maximum of the last ice age, but at its very end – that is, about 10,000 years ago. It has been suggested by P. Martin that the mammoth, the mastodon and the woolly rhinoceros became extinct at about that time because they were overhunted by man. This suggestion has the attraction that it distinguishes between the last ice age and all the preceding ones, which the animals survived. (It was only during the last ice age that man's hunting technology could have threatened so powerful a beast as the mammoth.) The suggestion only seems plausible, however, for regions where the human population density was reasonably large – for example, southern France. I doubt whether it can be considered plausible for so remote a place as the Leptev Sea, where the low precipitation characteristic of an ice age appears to have prevented the growth of an ice sheet. A region known as Beringia, in north-eastern Siberia, is known from plant and insect data to have remained ice-free, thus providing a habitat for animals such as the woolly mammoth that became adapted to a hardy climate.

Complete mammoths with surprisingly little degeneration of the flesh have been recovered from actual present-day ice in Siberia, while the bones of other animals have been recovered from near the shore of the Leptev Sea. It is hard to see how they could have become entombed in ice except at a time of violent change. A heavy downfall of freezing rain (i.e. rain falling as supercooled liquid which becomes ice immediately on impact with the ground) might have frozen them quickly enough to prevent decomposition. Since the mammoths survived a sequence of previous ice ages, it is unlikely that they were overwhelmed by normal cold or by blizzards or an ice sheet that grew only a few inches a year.

I have been informed that, today, when reindeer fall down crevasses in the Greenland ice, they are subsequently found to be in an unpleasantly putrefied condition. It seems that, no matter how cold the air is, the body heat of the dead animal is sufficient to promote bacterial decomposition. The Siberian mammoths, in spite of their much greater body weight, have not putrefied in the same way, which would support the suggestion that they were robbed of their body heat at an extremely rapid rate – much quicker than conduction into cold air could give. Either they died of hypothermia caused by freezing rain, or they blundered into bogs and pools of exceedingly cold water formed from melting permafrost. The first of these possibilities is more likely to happen at the beginning of an ice age, the second at the end. When one considers the effect on mammoths of sudden gloom and of sudden heat from a brassy sky caused by the absorptive particles thrown up by an iron meteorite, all the evidence falls into place. The frozen ground would soften and the mammoths would flounder. Frozen pools and lakes would partially melt. In the conditions of poor visibility, the mammoths and other animals would be quite likely to blunder to their deaths in the icy bogs. The bones of many animals have been found in the permafrost of Alaska today: bison, musk ox, moose, horse, lynx, caribou and even the ground squirrel. It would seem that none escaped *local* extinction. These other animals did not become globally extinct simply because their habitats extended outside the heavily blanketed local regions. For the woolly rhinoceros, the mastodon and the mammoth, however, there was no escape; their habitat lay entirely within the affected locality, which seems to have constituted essentially the whole of the northern polar regions of the Earth.

The most famous extinctions in the whole of the geological record may have been caused by a similar event. Every one of the large dinosaurs disappeared suddenly about 65 million years ago – long before the advent of man. Many suggestions have been put forward for the cause of these extinctions. A widely discussed favourite in recent years has been the theory that a large influx of cosmic rays from space affected the dinosaurs like the lethal flash of radiation from the explosion of a nuclear bomb.

Stars of large mass end their lives violently. They become unstable and explode like a nuclear bomb, emitting high-energy cosmic rays as well as a brilliant flash of visible light. For a few days after such an explosion, called a supernova, the exploding star is as bright as the whole of our Milky Way. Such explosions occur at random positions with respect to our solar system, about two or three times each century. Over the whole history of the Earth there must have been about 100 million supernovae. It is argued that the distance from the Sun of the nearest of these supernovae may have been no more than thirty light years (one light year is the distance travelled by light in a year – nearly 10 million million kilometres), and it is possible that the nearest supernova occurred by chance 65 million years ago.

The flood of cosmic rays from such a supernova might greatly exceed the normal rate of incidence of cosmic rays on the Earth – possibly a thousandfold or more. Since the normal annual rate at which cosmic rays from space impinge on the Earth's atmosphere is known to generate radiation that is about one ten-thousandth of the lethal dose for all land animals, the flood of cosmic rays from a supernova might conceivably have caused the animal extinctions through widespread radiation deaths.

A variant on this theory replaces the nearby supernova with an abnormal event on the Sun. The Sun emits a more or less continuous low flux of cosmic rays. Perhaps 65 million years ago the Sun emitted a much larger flux of cosmic rays, sufficient to induce widespread radiation deaths on the Earth. Why? Why not? Another variant of the theory argues that, instead of cosmic rays acting directly at ground level, they acted indirectly, largely destroying the layer of ozone high in the Earth's atmosphere, which normally prevents lethal solar ultraviolet light from reaching ground level. Again why? Or why not?

I will explain why not. Indeed, I doubt that the supporters of these astronomical theories would have bothered to suggest them if they have been aware of the full range of the extinctions that occurred 65 million years ago. The extinction of the dinosaurs has received so much publicity that one tends to concentrate on their part of the story. In fact, the dinosaurs were

161

only a rather minor part of the catastrophe. Every terrestrial animal weighing more than 50 pounds, whether a land animal or a fish in the sea, seems to have become extinct at that time. Even this still gives only a restricted impression of the disaster. More complete information has been set out in terms of numbers of genera for various kinds of organisms by D. A. Russell (K-Tec, *Syllogeus* No. 12, p. 14). The details are shown in Table 4.

Table 4: **Numbers of genera involved in the late Cretaceous extinctions**

	Before extinctions	*After extinctions*	
Freshwater organisms			
cartilaginous fishes	4	2	
bony fishes	11	7	
amphibians	9	10	
reptiles	12	16	
	36	**35**	(97%)
Terrestrial organisms			
(including freshwater organisms)			
higher plants	100	90	
snails	16	18	
bivalves	10	7	
cartilaginous fishes	4	2	
bony fishes	11	7	
amphibians	9	10	
reptiles	54	24	
mammals	22	25	
	226	**183**	(81%)
Floating marine micro-organisms			
acritarchs	28	10	
coccoliths	43	4	
dinoflagellates	57	43	
diatoms	10	10	
radiolarians	63	63	
foraminifers	18	3	
ostracods	79	40	
	298	**173**	(58%)

Bottom-dwelling marine organisms	Before extinctions	After extinctions	
calcareous algae	41	35	
sponges	261	81	
foraminifers	95	93	
corals	87	31	
bryozoans	337	204	
brachiopods	28	22	
snails	300	150	
bivalves	399	193	
barnacles	32	24	
malacostracans	69	52	
sea lilies	100	30	
echinoids	190	69	
asteroids	37	28	
	1976	**1012**	(51%)
Swimming marine organisms			
ammonites	34	0	
nautiloids	10	7	
belemnites	4	0	
cartilaginous fishes	70	50	
bony fishes	185	39	
reptiles	29	3	
	332	**99**	(30%)
Overall	**2868**	**1502**	(52%)

Coccoliths are classified botanically as photosynthetic algae. Their name comes from a tiny ring of calcium carbonate secreted in their cell membranes. Coccoliths are deposited on the sea floor in enormous numbers; annual rates up to 6000 million per square foot have been estimated – that is, more than one for every human being on earth.

Laboratory experiments show micro-organisms to be far more tolerant of high radiation doses than large multi-celled animals, which is understandable because there are many more vulnerable points that can be damaged in a large animal like a dinosaur. Yet the extinction of coccoliths – thirty-nine genera (related groups of species) out of an initial forty-three – was much more

163

severe than the extinction of the land-based reptiles (which included the dinosaurs) – thirty out of an initial fifty-four. This fact alone militates against the astronomical theories.

Water acts as a shielding agent against cosmic rays. Marine waters are in general deeper than freshwater systems, so that extinctions, if caused by cosmic rays, should on the whole have been worse in fresh water than in the sea. The examples of both reptiles and bony fish in Table 4 show the reverse. Indeed, the genera of freshwater reptiles actually managed to increase, as did the mammals, through what was otherwise a period of drastic extinctions. Altogether, extinctions were generally more widespread in the ocean than on the land, in clear contradiction to the radiation-dose theory.

Interruption of sunlight for a temporary period is the single factor that comes nearest to explaining the extinction data. Such an occurrence extending over a year or more (the time micropartices would take to fall down through the stratosphere) would create a food-chain problem for every animal. Photosynthesis, the necessary basis of all the life-forms mentioned in Table 4, would stop. Heavy animals are far more vulnerable than small ones, because their necessary minimum of food is larger. Even a single year would be quite long enough for every large browsing animal to die, and with their deaths the large carnivorous animals would also starve.

Food chains can be shorter in the sea than on the land. Small land animals and freshwater fish could in many cases subsist on nuts, seeds and the vegetation that would lie around for several years. Land plants would scarcely be affected either, because seeds and roots would re-establish the flora when sunlight eventually returned. The gravest problem would be for marine fish which feed either directly on photosynthetic plankton or are dependent on eating other creatures which do feed on these plankton. Such a chain would be broken almost instantaneously.

Photosynthetic plankton generate another, less direct, chain through the organic material which they make available to the non-photosynthetic plankton and to other creatures as well. The reservoir of this organic material would probably survive for a while. Marine creatures dependent on it would therefore have a better chance of avoiding extinction, rather like the small land-based creatures dependent on decaying vegetation. And, like land-based creatures that survived by eating buried seeds and nuts, there would be marine bottom-dwellers that continued by abstracting the dead plankton which had fallen over the previous millennia to join the ooze on the ocean floor.

Deprived of sunlight and nutrients, some forms of photosynthetic plankton living in shallow waters adopt a dormant condition. They fall to the sea floor, where they remain inert, perhaps for many years, until favourable conditions return. Such photosynthetic plankton, including the diatoms and dinoflagellates, had a ready-made method of survival. The situation for marine creatures was not therefore by any means hopeless, and about a half of the marine species managed to survive, as we can see from a glance at Table 4.

It is satisfactory that two apparent difficulties in these ideas are both resolved by the same aspect of our discussion of giant meteorites (Chapter 9). First let me present the difficulties.

We saw in Chapter 9 that Apollo-type objects, giant meteorites with diameters of a kilometre or more, hit the Earth every 100,000 years or so. As an outcome of these collisions, the cutting-off of sunlight must be a comparatively frequent event, which in general produces nothing like the biodisaster shown in Table 4. Evidently, there was something special about 65 million years ago.

Photosynthetic plankton are not adjusted to be most efficient in their conversion of carbon dioxide and water into sugars when exposed to the normal intensity of sunlight. Plankton attain their maximum efficiency at as little as one-tenth of the intensity of normal sunlight. Consequently, a doubling of the reflectivity of the Earth – say, from its present 36 per cent to 72 per cent (reducing the sunlight penetrating to the ocean surface from a present average of 64 per cent to an average of 28 per cent) – would have no more than a minor effect on the plankton. There would be no marine extinctions to compare with Table 4.

Photosynthetic plankton, kept in a warm and active condition, begin to respire when the light intensity falls below a few per cent of normal sunlight. They begin to live on the sugars which they have accumulated. The respiration causes the sugars to be converted back to carbon dioxide and water, by processes similar to

165

those which occur in other non-photosynthetic organisms. And, as with other creatures, the process ends in death when the sugars are all used up.

It follows that, to produce extinctions among photosynthetic plankton, the light intensity must fall to very low levels – considerably lower than we have contemplated in the previous chapters. The climatic balance affecting the occurrence of ice ages could be more than tipped by a halving of the incident sunlight. We are now concerned, however, with a very much more extreme reduction than this. Indeed, the reduction of sunlight necessary to explain Table 4 must be so drastic that it can hardly be explained at all in terms of an increase in the reflectivity of the Earth, but is consistent with *absorption* for a reason I will now explain in detail.

The difference between reflection and absorption is very obvious from simple everyday experiences. When a reflecting cloud crosses the Sun, the light level falls, but only by a stop or two on a camera. But, if you take a hollow tube and seal one end with a bit of sheet metal, you can point the tube straight at the Sun without any light at all reaching your eye at the open end. This is a case of absorption.

With particle blankets around the Earth the differences can be enormous. If a reflective blanket reduced the intensity of sunlight at ground level fifteenfold, the day would certainly be very gloomy, but because of the great range of adaptability of our eyes we would be able to see quite well, as we can just after the Sun has set. But, with an absorptive particle blanket of the same 'optical depth', as it is called, daylight would fall to about the level of a starlit night; increasing the thickness of the absorptive blanket a little further would produce a complete blackout, as happened with the bit of sheet metal at the end of the tube.

There can also be a marked difference in the effects of reflection and absorption on the temperature of the Earth. A reflective particle blanket cools the Earth, whereas an absorptive one of fine particles of iron would warm the Earth. Particularly in the tropics at midday, such a blanket would produce a condition of truly oppressive heat. Literary writers have sometimes attempted to describe oppressive heat by referring to a 'blazing sun in a sky of brass'. With a highly absorptive blanket of small metallic particles, there would be no blazing sun, no visible light, but there would be something very much like a sky of brass, hot brass at midday. There would be huge evaporation from the ocean and rain, perhaps endless rain.

The giant meteorites of iron that were considered in Chapter 9 to hit the earth every 100,000 years were not remotely of the scale to cause the biodisaster of Table 4. The meteorites of Chapter 9 were only required to produce a 20 to 30 per cent absorption of sunlight, whereas the meteorite that could cut out all sunlight from the ground and produce the disaster of 65 million years ago would have had to be at least 10 kilometres in diameter. That disaster was therefore uncommon because it was caused by the collision of the Earth and a body exceptional both in scale and composition.

The extinction of the dinosaurs becomes easy to understand. They were unable to see in the gloom, and they were assailed by widespread flooding and starved by a chronic lack of food. Heavy rain and flooding are known to have occurred. David Attenborough wrote in *Life on Earth* of the extinction of the huge horned dinosaur Triceratops:

> Just above the level at which its most recent bones are found, a thin deposit of coal rules a black, precise line that can be traced in cliff after cliff across Montana and over the Canadian border in Alberta. It must represent short-lived but widespread swamp forest, and it marks the death of the dinosaurs. Immediately below it you can find the remains not only of Triceratops, but of at least ten other species of dinosaur. Above it are none.

One more point is worth making. Heavy metals can be very toxic. A huge metallic body colliding with the Earth could conceivably have augmented the short-term chaos which it produced by adding a longer term legacy of poisons. Some of the extinctions which occurred could have arisen from this legacy.

The above is a general description of technical papers that were cited earlier in this chapter. Similar ideas have been proposed very recently which prove beyond cavil that the ecodisaster of 65 million BP was associated with the presence of unusual material in the atmosphere of the Earth. W. Alvarez and D. W.

Vann (*Forum*, February 1979) have given arguments to suggest that a layer of unfossiliferous clay about 1 centimetre thick, discovered at Gubbio (not far from the scene of the Roman victory over Hannibal at Lake Trasimeno in the Umbrian region of central Italy), was laid down contemporaneously with the ecodisaster. This clay layer, and two others of very recent discovery, would seem to have a connection in time with the fine layer of coal in the cliffs of Montana and Alberta mentioned by David Attenborough.

The Gubbio clay layer was found to contain about one part per million of the precious metals iridium, platinum and osmium.[2,3] Such abundances are typical for cosmic material but they are several hundred times too large for normal clay, or indeed for the rocks of the Earth's crust generally. The inference is that the Gubbio clay had a cosmic content – perhaps essentially the whole of it was cosmic.

Alvarez and Vann consider the cosmic material to have arrived as a giant stone meteorite, considerably larger than a normal Apollo-type object. While one can distinguish the event of 65 million years ago in this way, rather than as a giant meteorite of metal, the above discussion suggests there may be difficulties in this change of emphasis. A giant stone meteorite would throw up a blanket of reflective, not absorptive, particles into the atmosphere, reducing the incidence of sunlight at ground level to a gloomy situation but not a blackout. Since photosynthetic plankton can survive and even remain active in a considerably weakened intensity of sunlight, the situation for the incidence of a stone meteorite does not seem as unequivocal as it would be for a meteorite of metal.

While an exceptionally large stone meteorite might prove to have been the cause of the ecodisaster of 65 million BP, I do not think stone can be substituted for metal in the case of the extinctions of the mammoth and mastodon. Nor would a giant stone meteorite, which gives reflection and cooling, bring an ice age to its end. A sudden warming was needed 10,000 years ago accompanied by inky blackness, thus causing the mammoths to plunge to disaster.

[2] L. W. Alvarez, W. Alvarez, F. Asaro and H. V. Michel, in *Science*, 208, 1980, 1095, [3] and R. Ganapatny, in *Science*, 209, 1980, 921.

11 Man Versus the Antarctic

I doubt that a repetition of any of the natural disasters we have considered above could result in the extinction of man – not even a disaster of the magnitude of 65 million years ago. With a total disappearance of the Sun, but without a collapse of the temperature, with a warm, literally metallic sky, there would be a comparatively easy survival for many people over the first few months. So long as fuel stocks could be maintained at the power stations there would be artificial light.

Food would be the overriding problem, just as it was 65 million years ago. It is easy to imagine the early frenzied rush to the stores, the impulse of everyone to hoard everything in sight. Distribution of food to the shops would not keep pace with demand. Nor could those modern governments that win electoral power by perpetually pandering to the demands of their people hope to control the situation. There would be early appeals by political leaders, appeals for everybody to remain calm, assurances that everything was in the government's good and capable hands. But the people would know perfectly well that the situation was entirely otherwise and nobody would remain calm.

By the time I was ten years old I had learned all about hoarding. In my home village the Sunday School held 'treats' for the local children from time to time. We were sat down at a long table on which there was an apparent multitude of plates of tarts and cakes. We were not permitted to make a beginning until a prayer had been said. This measure of religious discipline on our part seemed to satisfy the adults, who paid little attention to ensuing events. We made no attempt to gobble like a common animal, since gobbling is far too inefficient a technique to befit a creature so acutely intelligent as man. Within a minute of the word 'go' being given, however, not a particle of food was to be seen anywhere on the table. It was all carefully and wisely stored underneath. We did not discriminate. We acted with commend-able forethought, storing away both fancied and unfancied morsels. Thereafter it was a question of barter. The severe mauling which the various sweetmeats undoubtedly suffered in these processes of exchange never seemed to do much harm to their taste.

So it would be everywhere, but of course the 'treat' could not last. It is a sobering thought that man, for all his modern technical sophistication, is still almost entirely dependent on the coming harvest. Even the world's greatest food-growing areas keep no more than a two- or three-year stock of grain. Present-day society is totally vulnerable to a cutting-off, or even to a partial cutting-down, of sunlight over a ten-year period.

The bulk of mankind would gradually starve at different rates according to the amount of food that happened to be stored in each particular locality. Because of the dependence on artificial light, the various areas would soon become separated, with only poor communications even within nations, and essentially none between nations. The poorer nations which are at subsistence level would disappear in only a month or two. The richest nations could survive for perhaps a year. The first to go among the rich would be the health-food addicts, those who can nowadays be seen thronging stores cluttered with bins and bottles in uncanny resemblance to the shops of the medieval apothecaries.

The route to survival would lie in that *bête noire* of health-food addicts, tinned food. But not, of course, in the city communities, where every such tin would go into instant hoarding, like the tarts and cakes baked so many years ago by the well-meaning matrons of my home village. But somewhere, outside the pools of light which defined the cities, would be a Noah with his Ark, an Ark chock-a-block with tins of meat. Somewhere a commercial cannery would have established just such a store and somewhere the vicissitudes of communication would have left it untouched. It would be at the courtesy of the canning industry that mankind could survive an episode like the one which overtook the Earth 65 million years ago.

A better mode of survival would of course be to maintain a ten-year store of food, not just for Noah and his friends, but for everybody on the Earth. In the present-day condition of society this is hardly a feasible possibility. But, if man survives the present political and economic tensions of a grossly overpopulated world, there would be little technical difficulty in building such a store. Since the biodisaster of 65 million BP has happened only once in several hundred million years, it is a fair expectation that we have plenty of time in which to deal with our present social troubles, and to settle down into being a more thoughtful and far-sighted animal.

The case is otherwise with the threat of a new ice age. Here the problem may well arise on a timescale no greater than that which separates us from ancient Greece and Rome. There is absolutely nothing to be done about stopping future volcanoes and future impacts of giant meteorites with the Earth, nothing to be done about moderating the immediate damage such events might cause, nothing to be done about the throwing up of a reflective particle blanket around the Earth, and nothing to be done about ameliorating the following ten-year period of only partial sunshine. What can be done, however, is to stop the Earth from falling into a long-term pallid ice-age condition, an ice age which might very well last without hope of intermission over a period of 50,000 years or more. Such a long-term disaster can certainly be stopped.

The solution is to warm up the ocean to the condition it was in 20 million years ago. There were no ice ages then and, if the water in the deep ocean were again made as warm as it was then, there could be no ice ages in the future.

Where would the energy needed to warm so great a quantity of water come from? The answer is that it must come from the Sun itself; no such quantity could be supplied by man. Cold water pumped up from the ocean depth is to be warmed by the Sun when it reaches the surface of the sea.

Care would be needed not to pump up the cold water too quickly, otherwise the surface waters would be seriously lowered in temperature, thereby reducing the evaporation of water vapour and so weakening the heat engine which is now preventing the incidence of an ice age (see Chapter 7). Warming the ocean would put a further strain on the world heat engine, which, as we have seen, is none too powerful. Even so, it is reasonable to suppose that a 1 per cent strain could be tolerated without any serious consequence. Storing 1 per cent of sunlight in the heating

171

of ice-cold water pumped up from the ocean deep would increase the heat storage of the ocean by a whole year's supply of solar energy for each century of pumping. In a millennium, therefore, the heat storage would be increased from its present ten-year supply to a twenty-year supply. In 2000 years – the timespan that separates us from classical Greece and Rome – the heat storage could be increased further to a thirty-year supply of solar energy. At that stage, the Earth would probably have become safe from a further ice age. And, if the next giant meteorite did not come for another 4000 years, the oceanic heat storage could be increased still more, to an entirely safe margin with a fifty-year supply of solar energy, a supply comparable to that which existed 50 million years ago. Nothing then could destroy the equable climate of the Earth.

Because water in the deep ocean is denser than water at the surface, energy would be needed to raise the cold deep water. How in a society chronically short of energy could the pumps be driven? Miraculously, no external supply of energy would be required at all. The pumps could be driven by engines that worked on a cycle of heat transfer between the warm surface water and the cold water that was being continuously raised. Once the system had been set up, it would run itself.

The system could even be worked at some energy profit, and there are engineers who advocate it as an answer to the growing energy shortages in modern society. I do not myself believe these claims to be valid, because I do not think diffuse sources such as the ocean and the winds can ever give a satisfactory primary supply of energy. The primary supply must be highly concentrated in its form. Given a concentrated supply, however, energy from diffuse sources can often be useful in particular circumstances. There are situations in which a windmill can be useful, provided coal is available as a primary supply to manufacture the windmill. But no high-grade technology could ever be directly dependent on windmills, or on the pumping of cold water up from the deep ocean. Nevertheless, the excess energy obtained from oceanic pumping could certainly be used to help the system pay its way.

How large would the system need to be? If the whole ocean below 500 metres depth is to be warmed in, say, 3000 years, then about one three-thousandth part of the ocean must be raised each year. For a reasonable pumping speed of 1 metre per second, the time of ascent of water from the average oceanic depth of 3790 metres would be a little over an hour. Since there are rather more than 8760 hours in a year, the pumps would circulate water up and down nearly 9000 times per year. The area of the sea over which pumping would need to take place is now very easily estimated. It is just a fraction (1/3000 × 9000) of the whole ocean surface – equivalent to the area of a circle with a radius close to 2 kilometres. The undertaking would be substantial, but nothing to daunt a modern society which prides itself upon high technical competence. The necessary effort would not be greater, comparatively speaking, than the building of Stonehenge was to a neolithic people, or the pyramids to the ancient Egyptians, or the European road system to the Romans, or the first lunar landing to the Americans.

We saw in Chapter 6 that the present ice epoch was caused by the drift of the Antarctic continent towards the South Pole. In spite of the ocean's being very warm in the early stages of this drift, ice was already forming at 30 million BP and by 20 million BP ice sheets had spread across Antarctica. This fact of geological history is an assurance that deliberate action by man to increase the heat storage of the ocean would not cause the present Antarctic ice sheet to melt.

There would probably be some raising of sea level from a melting of Greenland ice, and indeed some partial melting from the Antarctic itself, but nothing like the 70 metre rise of sea level that would follow a total melting of all the Antarctic ice. In some places, sea walls would need to be built, but over 3000 years their building would hardly put much of a strain on the industrial capacity of society. The consequences would be quite minuscule compared to the devastating effect of a new ice age.

It is in any case likely that today's sea-covered continental shelves will not remain sea-covered for very much longer. The continental shelves are rapidly becoming of such great economic importance that their draining cannot be much more than a century away. At the present price of land in the British Isles, the cost of reclaiming potentially rich sediment-covered areas around the whole coastline is not far beyond what would even be

173

economic today. The time must surely be close when the British population will wake up to a sudden realization that with a little effort it would be possible almost to double the area of their country, and with land that would eventually be extremely fertile. For this reason, too, I do not think that on a timescale of many centuries there would be any serious worry if sea level were to rise by a few tens of metres. Long before this happened, the sea is likely to have been excluded from its present-day control over so much valuable land.

It is impressive to consider how much more effective the brain of man could be than the raw power of the Antarctic. The Antarctic took millions of years to cool the ocean. Man, if he wishes, could warm it in only a few thousand years.

Suppose we were permitted to enjoy a utopian world of health and youth, a world in which we grew from childhood until the early twenties and then never aged any further. Like the algae, bacteria and some fungi we would be potentially immortal. Yet, like a tree, we would die sooner or later from some chance happening.

Trees are much more at risk of death through being struck by lightning than we would be. In the British Isles, where fierce thunderstorms are less common than in some other countries, we could expect to survive for about 10 million years before being hit by lightning, or by bits of disintegrating aircraft. Domestic pets would be something of a threat, but it would also be 5 to 10 million years before we would die through a dog bite or a cat scratch. The risk of death through nuclear energy has been calculated to be about the same. Some people claim this calculated nuclear risk (giving us a life expectancy of 10 million years) is considerably too small, while others think it may be too large. Be that as it may, all these distant risks are much smaller than other risks we commonly accept without serious misgivings.

There is a risk as we walk beside a road that the driver of a passing car will lose control, and that the car will fly off the road. There is a similar risk that a car, lorry or bus will mount the pavement while we are walking the local high street. Statistics show that we could expect to live 10,000 years before being killed in such an event. Diseases from the invasion of our bodies by bacteria and viruses, of which influenza is the most common, would be an even more potent threat. Disease would be likely to kill us in about 3000 years.

These are all unavoidable risks. We could also choose to indulge in activities that exposed us to voluntary dangers, as for instance mountaineering. Climbing rocks would kill us in about 5000 years, as also would the persistent driving of a car. There are other voluntary activities that would kill us much more swiftly than the ascents of apparently dangerous mountains, however. The persistent riding of a motor cycle would kill us in only fifty years, while the presistent smoking of cigarettes would kill us in 100 years.

Assuming we had the elementrary good sense to avoid these latter absurdly high risks, our expected lifetime (granted the boon of indefinite good health) would be a few thousand years. Very strikingly, this is just the same as the risk we run from the next ice age. Together with disease, the next ice age ranks as the biggest danger to which we as individuals are exposed. The next ice age is not a specific problem of the distant future. The causative agent, the strike of a giant meteorite, could happen at any time.

The risk of the next ice age is not just the biggest of the risks that we run. It is a risk that would hopelessly compromise the future. Besides wiping out a considerable fraction of those now alive, it would leave a wan, grey future from which the survivors and their descendants could do nothing to escape. It would be a condition that might last 50,000 years or more, a future in which the prospects for mankind would be much less favourable than they are today. This is why our modern generation must take action to avoid catastrophe, an ultimate catastrophe besides which the problems that concern people, media and governments from day to day are quite trivial.

Technical Note 1
Radiocarbon Dating

The Sun and planets are embedded in a diffuse sea of highly energetic particles known as cosmic rays. Every second a few of these particles strike each square centimetre of the Earth's outer atmosphere. The central nuclei of atoms in the atmosphere are fragmented from time to time by collision with these incoming cosmic rays. Among the fragments are particles known as neutrons, which are quickly reabsorbed by the nuclei of ordinary nitrogen atoms, causing the nitrogen to disgorge a different particle – a proton – thereby changing nitrogen to carbon,

$$\text{Nitrogen} + \text{Neutron} \rightarrow \text{Carbon} + \text{Proton}$$

The carbon atoms so formed are different from the common kind of carbon, because the atoms, being derived from nitrogen, have nuclei each containing fourteen particles instead of the usual twelve. Physicists denote this difference by writing ^{12}C for common carbon and ^{14}C for the small quantity of carbon arising from the effect of cosmic rays.

Chemical processes do not take much notice of the difference between ^{12}C and ^{14}C. Nor do biological processes. When plants absorb carbon dioxide from the air, it matters very little whether the carbon in the carbon dioxide molecules is ^{12}C or ^{14}C. Consequently, a small fraction of the carbon present in plants is ^{14}C. Because animals live by eating plants, or by eating other animals who eat plants, the carbon in animals also contains a little ^{14}C. The bones of mammals and the skeletons of beetles thus contain ^{14}C.

Whereas ^{12}C in biological material always stays ^{12}C, the ^{14}C changes slowly back to nitrogen. It does so in an interesting and precise way. After 5700 years, one-half of the ^{14}C remains, the other half having changed back to nitrogen. After a further 5700 years – that is, after a total of 11,400 years – one-half of the remainder has changed to nitrogen, leaving one-quarter still as

177

^{14}C. After a third period of 5700 years, one-eighth of the original ^{14}C is left, and so on, in further characteristic time steps of 5700 years. This interval of 5700 years is known as the 'half-line' of the ^{14}C atoms.

The ratio of the number of ^{14}C atoms to the number of ^{12}C atoms in a piece of wood or in the fossil skeleton of a beetle thus decreases as time goes on, and it does so according to a definite rule. If we know what the ratio was initially, say by examining a modern living beetle, and if we measure what the ratio has become in a fossil skeleton, the age of the skeleton can readily be calculated.

This is the radioactive carbon method used for the dating of ancient objects and for dating the British beetles discussed in Chapter 1. There is a limitation on the method. The object or skeleton to be dated must not be too old, otherwise too few ^{14}C atoms will remain for their quantity to be measured with adequate accuracy. For this practical reason, the method is at present restricted to objects not more than about 50,000 years old.

Technical Note 2
The Earth's Magnetic Polarity

On the left of Figure 42, we see how a small compass needle would align itself at various positions near an ordinary magnet. At the right, we see how small iron-bearing particles, acting as tiny compass needles, actually align themselves in the field of a magnet. If one were to turn the magnet around, so that the poles marked N and S in the figure were interchanged, the compasses at left would reverse their directions, as would the iron-bearing particles at the right of the figure. Such an interchange is called a switch of 'polarity'.

These properties apply to the magnetic field of the Earth. Iron-bearing particles deposited on the ocean floor or on the land, and iron-bearing particles present when molten rock cools

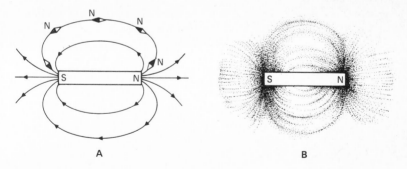

Figure 42: *Magnetic polarity. The magnetic field of the Earth causes a compass needle to become oriented in a similar way to the magnet in the left part of this figure. Small iron-bearing particles behave like compass needles, as do the iron filings shown in the right-hand part.*

and solidifies, all become aligned like small compasses with respect to the magnetic field of the Earth.

From time to time, at intervals of several hundred thousand years, the magnetic field of the Earth switches its polarity. (Why it does so has not yet been clearly explained.) Unlike a free small compass, iron-bearing particles bound tightly in deposits and in rocks may not be able to swing around when switches of polarity take place – they continue with their original orientations. But iron-bearing particles that are being newly deposited are indeed swung around to correspond with the new polarity. Thus iron-bearing particles are oriented in accordance with the polarity of the Earth as it was at the time when the particles were deposited. Cores of material, whether from the ocean or the land, will therefore show switches of orientation at various depths within them, namely at the depths which were being deposited at the times when the Earth's magnetic field made its switches of polarity.

No definite time measurements are yet involved, although fixed time markers are now set up. Thus, cores from widely different parts of the Earth can be compared, because the switches of orientation of the iron-bearing particles in each of them must have occurred at the same moments, namely at the particular times when the Earth's polarity made its switches.

If now for some quite independent reason any one of the cores can be dated, the time markers, applicable to all the other cores, become fixed. It may happen, for example, that a special core was deposited at a uniform known rate. Or one might have a core of so-called 'varved' material that was laid down in annual layers, in which case time intervals for the magnetic markers can be fixed simply by counting the number of annual layers between them. Varved material is found on the floors of freshwater lakes, usually in sub-Arctic regions where there are marked seasonal variations in the sedimentation rate.

The last occasion on which the Earth's magnetic field made a complete switch in polarity was at about 700,000 BP. Finer gradations of the time scale than this can be obtained, however. Jumps of the Earth's field occur more frequently than complete switches of polarity, and sudden jumps of direction can be detected as discontinuities at particular core depths, discontinuities in the orientations of iron-bearing particles. Jumps occur at intervals of about 30,000 years, so establishing time markers that are highly relevant for the ice-age problems discussed in the early chapters of this book, for instance the time intervals shown in Figure 6 (page 26).

The origin of the Earth's magnetic field probably involves phenomena at the boundary of the solid mantle and liquid core at a distance from the centre of about 2900 kilometres. This possibility may be relevant to the statistical analysis of ocean-core sediments (Chapter 4). Thus, the gravitational forces that cause both the variations in the tilt angle (Figure 18) and the precession of the rotation axis (Figure 17) act differently between mantle and core, which may conceivably introduce periodicities into the magnetic field which gain a spurious entry into the statistical analysis.

Thirty years or so ago, investigators of the problem of gravitational waves obtained solutions of Einstein's equation that were at first believed to be genuine physical waves, but which later turned out to be only waves that had been artificially introduced into the coordinate system. Likewise, certain of the periodicities obtained from the ocean-core sediments could be in the time coordinate, rather than in the ^{18}O content of the sediments.

Technical Note 3
Oxygen Isotope Analysis

Cores taken from the ocean bed can be analysed in a remarkable way, depending on what are called the 'isotopes' of oxygen. The three isotopes of oxygen are written as O–16, O–17 and O–18, with the nucleus of O–16 (the common form of oxygen) containing sixteen particles – eight protons and eight neutrons; with O–17 containing seventeen particles – eight protons and nine neutrons; and with O–18 containing eighteen particles – eight protons and ten neutrons. Isotopes have the same number of protons but different numbers of neutrons.

O–17 is too uncommon to be relevant to this discussion, so it will be convenient to omit it henceforth. Molecules of water are each made up of two atoms of hydrogen and one of oxygen, written as H_2O. Some of the oxygen in H_2O is O–18, about 0.2 per cent, and the rest is essentially all O–16. Now O–18 is heavier than O–16, and so the 0.2 per cent of water molecules that contain O–18 are heavier than the 99.8 per cent that contain O–16. Being heavier makes the molecules with O–18 fractionally harder to evaporate from liquid to vapour. This means that water vapour evaporating from the sea is deficient compared to the sea itself in molecules containing O–18. Rain and snow, and ice formed from the snow, are therefore deficient in O–18. It follows that, as an ice age develops, and more and more ice is deposited on the land, the concentration of O–18 relative to O–16 must rise in the water that remains in the sea (in order to compensate for the increasing deficit of O–18 in the ice on the land). Thus, the ratio of O–18 to O–16 in seawater rises as an ice age becomes more intense, and the ratio falls at the end of an ice age when melting of landlocked ice releases water (deficient in O–18) that flows in streams and rivers back into the sea.

182

These considerations show that, simply from an experimental determination of the ratio of O–18 to O–16 in seawater, we have a measure of how much ice there has to be on the land. An experimental determination for actual seawater can of course refer only to the present-day state of affairs, whereas what we really want to know is how much ice there was on the land in ages past. For this, we need a material that retains a memory in its composition of what the seawater ratio of O–18 to O–16 used to be at a time in the past. Sea animals with shells can do precisely this, since the calcium carbonate in the shells contains concentrations of O–18 and O–16 that reflect the seawater composition at the time when the animals were alive. When the animals died, their shells fell to the ocean bed and became enclosed in the sediments that were forming there. Since the sediments can be recovered from cores obtained in the ocean bed, modern measurements of what the amount of ice on the land used to be in times long gone by can be made.

There are two potential snags in this otherwise powerful method. Evaporation does not occur uniformly over the whole ocean, so that the concentrating effect of evaporation on the ratio of O–18 to O–16 must necessarily be variable, being greater in the tropics, where evaporation is large, than in high latitudes, where evaporation is small. Ocean currents have a strong stirring effect, however, and provided a few thousand years are allowed for mixing to take place this complication does not seem serious. The main effect is to smooth the record over timescales of a few millennia – we cannot expect to obtain the fine detail that was given by British beetles, as described in Chapter 1.

A second difficulty has caused some controversy. While the ratio of O–18 to O–16 deposited in the shell of a sea animal depends on the ratio present in the ocean during the life of the animal, it also depends on the temperature of the water. No clear-cut inferences can therefore be made from shell measurements of the O–18 to O–16 ratio unless either the temperature was effectively invariable over the past ages or the effects of variable temperature can be correctly allowed for in the determinations. Those scientists who have used the method sought the first of these escape routes. By choosing an animal that lives at some depth below the sea surface, where temperature variations

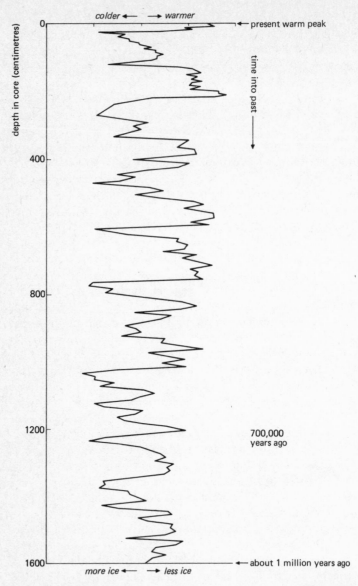

Figure 43: *Variation of Pacific sea temperature over time, as indicated by Pacific deep-sea core V28-238. (After N. J. Shackleton and N. D. Opdyke,* Quarternary Research, *vol. 3.)*

are thought to be small, the difficulty is believed to have been overcome.

Figure 43 shows results obtained by N. J. Shackleton and N. D. Opdyke from a Pacific deep-sea core (V28-238).

Technical Note 4
Electromagnetic Radiation

If one stands near the edge of the sea on an open beach, the ocean waves appear to be perpetually advancing, perpetually bearing down on you. Yet you can stand there quite dry. Although the waves are advancing, the water itself is not. Apart from the slow rise and fall of the tide, the water stays where it is, at unchanging distances from the shore.

The crucial properties of a wave, whether a waterwave or light or infrared radiation, are illustrated in Figure 44. In the case of the waterwave, the heavy dot at the left of the figure is a float which bobs up and down as the wave-form passes it by. The wave-form moves to the right between adjacent sections by one division of the imaginary ruler drawn on each of the sections. The distance between one wave-crest and the next is four divisions on the ruler, and it is this distance that is called the *wavelength* – λ say – that is, $\lambda = 4$ units of measurement. If the speed with which the wave-form travels from left to right is written as V, then the time interval between one section of the figure and the next is $I \div V$. The time for the float to make a complete vertical oscillation corresponds to four such intervals, $4 \div V$, which is the same as $\lambda \div V$. The reciprocal of this oscillation time, namely, $V \div \lambda$, is called the *frequency* of the wave, ν say. Thus $\nu = V \div \lambda$, or $\lambda\nu = V$. The frequency of the wave is the number of oscillations which the float performs in a unit time interval.

A similar relation holds good for lightwaves, or indeed for radiation of any frequency, $c = \lambda\nu$, it being usual to denote the speed with which the wave-form of radiation advances by c rather than by V. Since the value of c is accurately known, $c = 2.997929 \times 10^{10}$ centimetres per second (299792.9 kilometres per second), we evidently can determine λ if we know ν, and *vice*

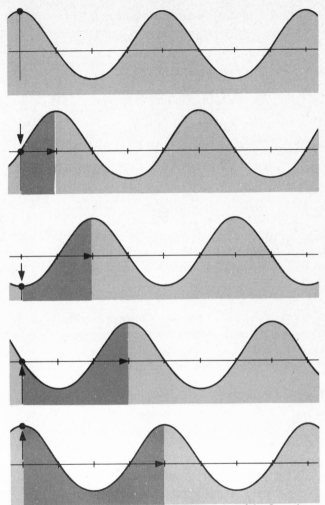

Figure 44: *Wavelength and oscillation. Movement of the float shows that the oscillation at any one point is completed in the same time as the whole wave takes to move through a distance equal to its wavelength.*

versa. Yellow light has a wavelength of about one two-thousandth part of a millimetre, or $\lambda = 5 \times 10^{-5}$ cm.

The corresponding frequency is about 6×10^{14} oscillations per

187

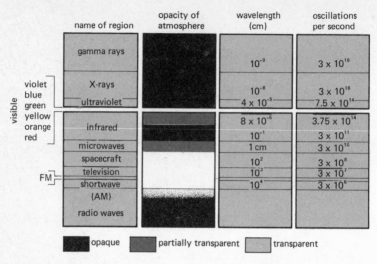

Figure 45: *The electromagnetic spectrum, and the opacity of the Earth's atmosphere.*

second. Figure 45 shows the various ways in which radiation is described, together with the appropriate ranges of wavelength and of the oscillation frequency. The degree of opacity of the Earth's atmosphere to the various types is also indicated in a general way.

When you tune the dial of a radio receiver, you are picking out the radiation of a particular wavelength from many wavelengths that exist all around the receiver. Likewise, when you change the channel of a TV set, it is the wavelength accepted by the set that is altered. All the channels are of course in the room all the time. Likewise, there is light and infrared radiation around us all through the day. At night most of the light is gone, but the infrared remains. Each wavelength defines its own radiation, and all the wavelengths can coexist together, behaving independently of each other.

Index

190